Ken Williams

Gödel Forever

Through 90 Years of Foundational Claims

Ken Williams

GÖDEL FOREVER

Through 90 Years of Foundational Claims

Bibliographic information published by the Deutsche Nationalbibliothek

Die Deutsche Nationalbibliothek lists this publication in the Deutsche Nationalbibliografie; detailed bibliographic data are available in the Internet at http://dnb.d-nb.de.

Bibliografische Information der Deutschen Nationalbibliothek

Die Deutsche Nationalbibliothek verzeichnet diese Publikation in der Deutschen Nationalbibliografie; detaillierte bibliografische Daten sind im Internet über http://dnb.d-nb.de abrufbar.

ISBN-13: 978-3-8382-1786-4

© *ibidem*-Verlag, Stuttgart 2023

Printed in the United States of America

To the memory of Mrs. B.T. Lallas, the 10th grade Ensley High School Algebra-2 teacher who one day set aside her lesson-plan to clarify an arithmetical homework locution that no one was able to sort, neither at home nor during the entire class period that followed.

"Is," she announced, raising a finger when the bell rang signaling the end of the class, "means equals."

Table of Contents

Introduction

Nothing in the known history of mathematics is quite like the Incompleteness of Arithmetic (GI), either in substance, in the circumstances around its discovery, or in the effect it has had on mathematical thinking. Unlike many others, the Incompleteness revolution was not the culmination of a group or industry effort that began a new School (like Newton's Calculus or Frege's Logicism), but came very much out of the blue from the single-mindedness of Kurt Gödel. Its effect has been negative, too, decimating two of the leading mathematical schools of the day, with its academic research a professional cul-de-sac (Girard 2011).

It is remarkable that on this 90th anniversary, and after countless learned articles and entire texts on the subject, a leading authority can still write in good conscience that, "As regards Gödel's First Incompleteness Theorem and the matter of its proof, Gödel's own paper has yet to be improved upon" (Smorynski 2009, 122). Indeed, the simple irrefutability of Gödel's original presentation played a large part of its initial appeal. Take the Chinese Remainder theorem; while unknown to most new to Gödel's derivations, it is a tidbit, among others, that the unfamiliar upper-level undergraduate student will easily pick up on along the way.

This is a story of a proofing and its influence over ninety years that will unfold in four acts. We begin with the GI derivation itself, expanded over details that for lack of necessity at the time, or space, Gödel did not include in the original production. By the derivation we refer solely to Gödel's original. This is also a story of the peculiar regularity with which GI citations turn up in the literature of 20th century philosophy. Such literature subjects include the limits to machine intelligence, the redundancy of truth, and number ontology. For the first, there is the indirect self-referencing of the Gödel sentence, g, that presumably evades machine proof; for the other two, the universal generalization in its construction. It is only the well-put and narrowly detailed former story, we think, that properly informs our understanding of the latter, broader story, bringing us up to where we find ourselves in it today. If there is one thing we have learned in the course of this study, it is that the facts regarding GI's proper place in these matters lies in the mathematical details of its derivation.

There are statements *in* arithmetic that cannot be shown either true or false *by* arithmetic. If the arithmetic statement "3+5 = 8" can be shown, i.e., proven true or false, it would seem then that "3+5 = 8 and/or 7+1 = 3" could also be proven true or false by the same schoolhouse arithmetic. But for the statement referred to in the Incompleteness claim (presumably, some conjoined and disjoined combination of like simple arithmetic statements) apparently the reasoning does not hold. How could this be?

There may be no answer nor need for one. Besides the Incompleteness claim, there are other interpretations of Kurt Gödel's formal deduction, the Gödel result (GI),[1] that range from the banal acknowledgement that consistent axiomatic formulations of number theory include undecidable propositions, to those that have it that the human mind is unable to comprehend itself, and by which computers can never outsmart humans (Jones and Wilson 2009).[2]

Here are a few others:

- All mathematics can be formalized: however, mathematics can *never* be exhausted in *any one* system but requires an infinite sequence of discourses which get progressively more comprehensive (Carnap 1934).
- Every arithmetic is incomplete (Waismann 2003).
- Truth transcends proof (Vidal-Rosset 2006).
- Every system of arithmetic contains arithmetical propositions which can neither be proved nor disproved within the system (Gödel 1962).
- An axiomatic approach to number theory cannot fully characterize the nature of number-theoretical truth; what we know and understand about mathematics transcends what can be expressed through our mathematical systems (Lipscomb 2010).

---

[1]   Bertrand Russell, philosopher and author of *Principia Mathematica*, which Gödel calls out by name (Whitehead and Russell 1910), seems to have interpreted Incompleteness as a verdict on mathematical consistency and openly worried whether 3+5 indeed *is* 8. There is no record he ever advocates we stop teaching it in schools.

[2]   Both seemingly at odds with a reasoning from GI that a mind may *not* claim superiority over a machine (Makey 1995). For others still, GI settles the question whether a machine has a soul (A. Turing 1950).

- Relying on words to lead you to the truth is like relying on an incomplete formal system to lead you to the truth. A formal system will give you some truths, but a formal system, no matter how powerful, cannot lead to all truths (Hofstadter 2000).
- In any language there exist true but unprovable statements (Uspenski 1989).

The list may be continued.[3]

Once the *pons asinorum* of *Mathematical Logic* (Lucas 2002), a new trend describes GI as "so simple, and so sneaky, that it is almost embarrassing to relate" (Rucker 2008, 162), with the details unnecessary to contemplate.[4] Here we maintain that even as valid interpretations may differ, the sound interpretation of Incompleteness is based on knowledge of precisely what it is Gödel has done. Common misconceptions of Gödel Incompleteness (GI) discussed in the coming "Indirect Self-Reference" section make this point clear. While covering the Incompleteness derivation in good detail, we do not run the full circuit. At one point in the original derivation, Gödel launches a sequence of 46 functional definitions essential to the derivation. It is tedious and takes up a lot of space. Of the 46, we analyze an essential few and describe others, leaving the rest for review in cited sources. Among these we recommend a modern translation of the original paper, *On Formally Undecidable Propositions of Principia Mathematica and Related Systems* (Gödel 1962), a popular review (Smith 2013), and a recent online rewrite (Gödel 2000) from which we freely borrow notation.

Of the differences here with these and other reviews there is our emphasis on the question of expressibility of primitive recursion in the formal language of arithmetic. That the Gödel sentence, g, appears there, as required for its Incompleteness, is a straightforward but lengthy demonstration usually taken as given. Another is that we make more explicit Gödel's original "arithmetization" of the meta-language, as consequent on an isomorphism between the original language, P, and another, P' that has a meta-language all its own. This helps draw out more clearly the underlying premises involved in the Incompleteness mechanics.

---

[3]   See e.g. chapter 11 of Sokal's *Fashionable Nonsense* (Sokal and Bricmont 1998).

[4]   "Understanding Gödel isn't about following his formal proof, which would make a mockery of everything Gödel was up to" (Jones 2008).

As will become apparent by the "Correspondence"–sub and "Indirect Self-Reference" sections below, it also helps clarify an important limitation on inferences from Incompleteness that might be made in other fields of study, such as the Philosophy of Mind that typically follow from the misconception of Gödel's arithmetization as an isomorphism between mathematical P and its meta-language, meta-P. If there is an isomorphism at play, and only two languages, then the isomorphism can only be between those two, inviting such speculations as exceed that limit; if there is but one reason for fleshing out this four-language, indirect self-referencing mechanism that infers "truth" without "meaning" and upon which GI proceeds, that is it. While there are two Incompleteness theorems derived in (Gödel 1962), the first and second, there are also two distinct proofs of the first—a semantic and a syntactic proof. In our first two sections we consider only the semantic derivation of the first, said to be the weaker proof.

Our first three sections are: "I. Arithmetization", "II. Indirect Self-Reference", and "III. An Odious Turn". Given the four flavors of Gödelian Incompleteness:

- Semantic Incompleteness, semantically derived (applying the notion of truth)
- Semantic Incompleteness, syntactically derived (absent the notion of truth)
- Syntactic Incompleteness (Undecidability), semantically derived
- Syntactic Incompleteness (Undecidability), syntactically derived

When the subject in question does not concern formal soundness itself, one or other of the simpler semantical derivations is usually cited; otherwise not. Accordingly, our beginning "I. Arithmetization" section is an elementary examination of the simplest semantic GI derivational details. While it is well known that the means by which g is said to indirectly refer to itself is established via the "arithmetization of a metalanguage" whose object is isomorphic to the language in which g appears, the crucial question becomes *where* precisely in this mechanism is the much-talked-about linguistic isomorphism applied,[5] and where not. The section closes out with the following Indirect Self-Reference inference diagram

---

[5]   Alternate interpretation, translation, what have you.

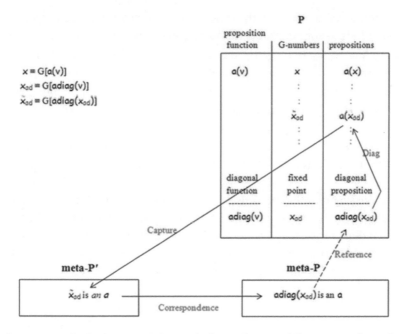

$$x = G[a(v)]$$
$$x_{od} = G[adiag(v)]$$
$$\tilde{x}_{od} = G[adiag(x_{od})]$$

by means of which toward the end of our short middle section, "II. Indirect Self-Reference", we closely examine how GI is cited in the context of the machine-intelligence limit discussion and consider whether, as usually assumed, g is ambiguous between object- and meta-linguistic readings.

Any presentation of Gödel's Incompleteness that aims at clarity must balance between deductive detail and economy of exposition, particularly with regard to details of its more novel features such as the formal "Definability" of primitive recursion relations and their "Diagonal Compositions" that drive the indirect self-referencing mechanism and is common to all four flavors. On the other hand, not only g, but all of the captured diagonally-composed metalinguistic predicates are locked into this mechanism, so that even the most elementary of these puts its inner workings on full display, as in the above inference diagram for arbitrary predicate "a". For these reasons, the 'Definability' and "Diagonal Composition" of primitive recursion in our formal system P, as well as P's Capture of its most elementary linguistic syntax "a is a variable", we derive in great detail, with examples and visual aids that together leave little room for misunderstanding. On the other hand, a comparatively detailed Capture of P's deductive syntax we do not give; such a treatment up to Gödel's final highly synthesized '𝑎 is not a theorem of P' would occupy several additional pages while revealing nothing more on the Incompleteness structure. For expository detail against economy, this combination strikes (we think) an optimal pedagogical balance found nowhere else in the literature. Readers familiar with the self-reference mechanism might initially want to skip *I. Arithmetization* and in the course of reading on refer back to it (as indeed we here will often do by cross-reference) as needed.

In section "III. An Odious Turn" we look at how GI citations show up in truth-theory literature, in particular those concerning the question of the redundancy vs transcendence of truth. Before rehearsing Gödel's syntactic derivations, however, we briefly detour to Kongsberg of 1931 where Gödel first announced his finds. Only there, GI shows up as corollary to his argument against the sufficiency of Hilbertian adequacy—that consistency is insufficient to guarantee existence—his main object. This might better explain why he goes to the trouble to avoid the semantical derivation, taking instead a more difficult syntactic route instead: a better, more tangible explanation than the conventional one, if not entirely in line with Gödel's own offered years and decades later.

As regards the appearance of GI in an extensive literature that takes it as mathematical proof of the transcendence of truth over proof, we simply recount Feferman's rather simple redundancy solution (there called "reflection") to Shapiro's transcendence account (there called "thick"), adding much needed detail, clarity and context. Its final sub-section, "Lisbon & Pisa, 2014" we reserve for special consideration of a relatively recent attempt to insinuate GI (via the universal generalization that appears in Gödel's ω-consistency) into a long-running Realist/Anti-Realist number-ontology dispute. The article we single out also dismisses Feferman's aforementioned answer and among other things makes the deflationist an anti-realist. We offer a scathing review.

Our final section "IV. Wittgenstein's Perspective" was conceived months after completion of all the others, coming as an afterthought. Clearly something is missing when the leading philosopher of language of the 20th century is not given a say on its mathematical magnum opus, particularly when half his writings around the time of its production (circa

1929-1944) had been devoted to the subject. Ludwig Wittgenstein was reticent on GI as on all other subjects, and such writings as those mentioned remained unpublished during his lifetime. What has since come down to us are mere fragments. They do, however, manage to reveal Wittgenstein's keen interest in, if criticism of, GI—these comments were contained in two or three unpublished notebooks from that period and which finally came out as part of his Nachlass collection of notebooks, manuscripts, typescripts and dictations in the late 1950s. We consider here a couple key moments in the corresponding secondary literature, that not unlike that of Gödel's Incompleteness, has taken on a self-generating, enduring existence all its own.

I. Arithmetization of Metamathematics
(PRR $\Rightarrow$ P $\Leftrightarrow$ meta-P' $\Leftrightarrow$ meta-P $\rightarrow$ P$_{fp}$ $\Leftrightarrow$ P)

To the interested student, the first challenge would seem to be to make sense of the Incompleteness claim itself, whose intended meaning is by no means immediate to the non-specialist. Indeed, any such specialist living before 1910 would scarcely have known what to make of it. At the turn of the 20th century, a phrase "statement of arithmetic" would have still been defined by its sense, understood to designate subject matter, much as a "statement of physics" still does.[6] Who then could have guessed that

$$\forall \varphi_1 [\forall x \, \forall y \, (\varphi_1(x,0) = x \cap \varphi_1(x,sy) = s\varphi_1(x,y)) \rightarrow \varphi_1(ss0,ss0) = ssss0]$$

would within 20 years be accepted as the rigorously correct form that expresses the sense of 2+2=4, while the form "2+2=4" itself would not show up.[7] As we shall see, the modern "statement of arithmetic" is a formula that conforms to a set of formation rules over an alphabet: a well-formed formula.

The concept "provability" has also undergone significant change, more formalized now than ever. The definite sense it has today, at least among mathematicians and philosophers of mathematics, is actually incidental to the program (long running by the 1900s) to regiment logic into a style of mathematics. George Boole, father of symbolic logic, was one of its early contributors. Half a century later, the father of modern formal logic, Gottlob Frege, turned the program on its head, taking arithmetic not only as a "point of departure" for his logical calculus (Frege 1879, 8), but as a first-test case of its applicability. Starting out as no more than a curiosity (Benacerraf 1981), Frege's detour proved fraught with difficulty, beginning notably with the Russell Paradox, soon to be followed by others. As problems mounted, the sentiment among philosophers of mathematics grew that the solution lay in more rigorous symbolism, and nowhere could be found a more thorough treatment than in Russell and

---

[6]   As it happens, the argument for an incompleteness to Quantum Theory by Gödel's great colleague et.al. (Einstein, Podolsky and Rosen 1935) runs precisely from this criteria.

[7]   See e.g. section 5.3.2 of (Shapiro, Foundations without Foundationalism 1991).

Whitehead's *Principia Mathematica* (Whitehead and Russell 1910) (PM)—2000 pages and three massive tomes published in 1910.

Even so, the formalization of mathematics as known today only began in earnest after 1931 (post-GI). The curiosity then touted as Logicism had expanded its reach under the guidance of Russell and others, while the first test case remained unresolved, many would say, until Gödel, whose Incompleteness brought into sharp focus the need for an even more stringent formalization of arithmetic—that of the language itself. At one point (Gödel 1962), Gödel points out the absence of an explicit syntax in PM, and quite on his own, sets forth the appropriate alphabet and formation rules. It was this observation that gave crucial impetus to the subsequent developments of Gentzen, Church, Tarski and others toward the formulation of formal systems of mathematics in the specific form in which we see them today (Suber 2002).

PM's Formal System P

Half a century before Gödel, Italian mathematician Giuseppe Peano set forth a formal system of arithmetic axioms, taking zero and numerical succession as primitive concepts in no need of explanation:

- There is no natural number whose successor is zero.
- Any numbers with the same successor are identical.
- What is true of zero, and is true of the successor of an arbitrary number on the condition that it is true also of that number, is true then of all numbers.

These statements along with $3+5 = 8$ and other notions of natural quantity may be rendered in a language with the alphabet

$$\{0, s, n, m\ f_2, f_3, f_4, ..., (\ ,), \cup, \forall, \sim \}$$

where n and m are variables of type one, s is the successor function, $sn = n+1$, f_i is a variable or type i, and the other symbols carry their usual syntactic and logical meanings. We use the Comic Sans MS typeface throughout for its elements. The syntax, or formation rules, for the language are such that combinations of the form

$$0, s0, ss0, ...\ \text{etc.}$$

are numerals (or number signs—signs of type one), and equality between numbers is defined by predication

$$n = m \quad \equiv \quad \forall f_2 \,[f_2(n) \leftrightarrow f_2(m)\,]$$

where the shorthand "$\leftrightarrow$" is unpacked

$$(a \rightarrow b) \cap (b \rightarrow a) \quad \equiv \quad a \leftrightarrow b$$

with "$\rightarrow$" given by

$$a \rightarrow b \equiv (\neg a) \cup b$$

and logical conjunction "$\cap$" rendered as

$$a \cap b \equiv \neg\,(\neg a) \cup (\neg b)$$

We use the symbolism "$\equiv$" for the formal definitional relation and "$\rightarrow$" for material implication.[8] Peano's axioms are then cast

- $\forall n.\, \neg\, (sn = 0)$
- $\forall n\, \forall m\,.\,(sn = sm\,) \leftrightarrow n = m$
- $\forall f_2\,[f_2\,(0) \cap \forall n\,[f_2(n) \rightarrow f_2(sn)] \rightarrow \forall m\, f_2(m)\,]$

and the sense of 3+5 = 8 captured as

$$sss(sssss0) = sssssss0$$

More generally, alphabet strings of the form

$$n,\ sn,\ ssn,\ \text{etc,}$$

are to be called signs of type one, whereas a sign of type i > 1 is simply a variable of type i, f_i, where "i" is said to be its "order". For representation of numerical propositions as n-ary relations between numbers, we define the set of formulas. Elementary formulas are alphabet strings of the form a(b), where a and b are signs, b being one order less than a. The class of formulas is then defined to be the class that includes all elementary formulas, and all that satisfy the further formation rules that if a and b are formulas, then so are

---

[8]   Metalinguistic material implication will be distinguished by a double lined arrow, $\Rightarrow$.

$\neg\, a$, $a \cup b$, and $\forall x\, (a)$.[9] Such expressions that conform to this syntax are grammatically correct, well-formed parts of the language and by convention are said to be well-formed formulas [formulas]. Symbols [the alphabet], signs, and formulas: these are the important components of the language, while Peano's axioms above, selected from the formulas, and proofs—sequences of formulas that begin with axioms and proceed via deduction rules to a final "proven" formula—comprise the theory of our modern formal system P.

On the other hand, the formation and deduction rules are part of a language that talks about P and its meta-language, meta-P.

---

[9]   We will usually use a, b, c, …. as variables, and the notation $a(v)$ etc., only when we want to make the v dependence of a explicit.

P

alphabet	signs	formulas	axioms	proofs
0	:	:	:	:
s	n	:	:	:
n	:	:	$\forall n\neg(sn=0)$	:
m	:	$f_3(f_2(n))$	:	-----------
{f$_i$}	ssss0	:	:	:
)	:	:	:	:
(	:	$f_2(0)\cup f_2(sm)$	:	:
∪	f$_8$	:		-----------
∀	:	:		$\forall n\neg(sn=0)$
¬	:			:
	:			:
	:			:
				$sss(sssss0) =$
				$ssssssss0$

				:
				:

lexical rules →	sentential rules →	theory declaration →	deduction rules →
0,s0,ss0,... etc. are signs. : : :	: : a, b formulas $\Rightarrow$ $a\cup b$ formula : :	Peano Axioms	modus ponens

Meta-P

This is where Gödel takes it up and immediately veers through a lengthy digression. The Incompleteness claim is of course not a formula in P. Likewise, questions about the P-alphabet, P-signs and so forth, like formation and deduction rules, are all meta-P statements. As it happens, the Incompleteness claim follows informally from an antecedent meta-statement that we have been calling GI

$$g \Leftrightarrow \text{'}g \text{ is not provable'}$$

that if true i.e., if the sentence *g* exists in P—can be derived in P—leads immediately to the Incompleteness claim. Such a sentence would have unusual properties: on the one hand, the semantics (and syntax) of a P-statement; on the other, a syntax that binds it in correspondence to a meta-P statement. To derive it, Gödel first translates P to another language P' by means of his G-transformation (shown below) on its alphabet $\{0, s, n, m\ f_2, f_3, f_4,...)$, $(, \cup, \forall, \neg \}$ and formation rules, such that syntactic statements in its metalanguage, meta-P', are in correspondence with statements in P. This is known as Gödel's arithmetization of meta-mathematics. Thus, statements in meta-P' will be about what in P' are numerals, and correspond to P-statements, which are about numbers.

As the syntax of P is given recursively, the meta-P question, whether a given expression is, for example, a formula, is the question whether the expression can be traced by a kind of reverse recursion to some well-formed compound of elementary formulas. One might guess then that a *recursive* functional form would serve most conveniently and be transparent-to-interpretation for corresponding meta-P' statements, so that a proposition regarding the proper formation of P' signs might, with a little ingenuity, be captured as a functional recursive proposition about numbers. For example, a recursive propositional function over natural numbers, "isVar(x) ", is shown below to stand in a correspondence with the meta-P' proposition: "x is a P' variable".

The correspondence, PRR $\Leftrightarrow$ P $\Leftrightarrow$ meta-P', that appears as a segment of our subtitle we next achieve in three steps: defining the primitive recursion relation (PRR), establishing their representation in P, then establishing their correspondence with meta-P' propositions.

The PRR

The factorial is a classic recursively-defined function, given in a form reminiscent of a programming algorithm

$$0! = 1$$
$$(x+1)! = (x!) \cdot (x+1)$$

in which defining functions are the constant "1", the addition operator "+", and the multiplication operator "·", though it is given equally well in terms of the primitive notions of P: the succession function "s" and constant "0"

0! = s0

(sx)! = (x!)·(sx)

(assuming that the multiplication operator can also be rendered in terms of s and 0) which, for the purposes of provability and eventual definability in system P, is how we should like it; for the extent to which a function is recursively definable in P is that to which it may be computed in P. This prompts the definition of a primitive recursive function class (PRF), whose members are (1) functions corresponding to the primitive notions of system P: s and 0, (2) all functions given by composition of PRFs, and (3) all functions recursively defined by PRFs. Thus:

- s and 0 are PRFs
- if g(x) and h(x) are PRF, then f(x) formally defined as the composition g(h(x)) is also a PRF
- if g(x) and h(x, y) are PRFs, then f(x) recursively defined as
  f(0) = g(0)
  f(sx) = h(x, f(x))
    is also a PRF.

The PRF "f" is "primitive" in the sense that its defining functions (definiens) g and h are one definitional step nearer the arithmetic primitives, s and 0. The virtue of these particular two is their immediate computability.[10]

Thus, addition and multiplication defined by

x+0 = x

x+sy = s(x+y)

and

x·0 = 0

x·(sy) = x + x·y

---

[10]   Given that there is nothing *recursive* to the evaluation of compositions, and for other reasons (Soare 1996), the PRF might more intensionally have been designated "Algorithmically Computable Functions".

are PRFs, as then is the factorial function, each being traceable by a sequence of recursion definitions to the primitive successor and zero functions.

Definition

PRR ⇒ P

One might also define the factorial in the more familiar notation

$$n! = 1 \cdot 2 \cdot 3 \cdot \ldots \cdot (n-1) \cdot n$$

along with any number of others. It remains, in the capacity of a mapping, the factorial function, but not in PRF form. With the convenience of its ready correspondence with meta-P' propositions comes the inconvenience that the PRF has no means to explicit definition in the syntax of P, given for example its Spartan alphabet in which symbols for neither equality, multiplication nor addition appear. If g is to be derived, however, some manner of PRF definition in P will have to be made. The statement g will both correspond to a meta-P statement *about* the syntax of P, and be *in* the syntax of P; the first fact moves us to derive g as a PRF (given the recursive syntax of P), the second to attempt a definition of this PRF in the syntax of P, a feat tantamount to defining "n!" by some combination of symbols {0, s, n, m f_2, f_3, f_4,...), (, ∪, ∀, ~ }.[11]

Gödel might have gotten around at least part of this difficulty by means of PM's method of formal definition, as above for "equality", or he could have simply enlarged the alphabet to include "=". He chose neither—perhaps for economy or aesthetics (or esthetic economy), perhaps with an eye looking forward to some particular in the derivation, or perhaps to remain more faithful to the PM Logicist program that wherever possible eliminates non-logical symbols in favor of logical ones.[12] In any

---

[11]   Julia Robinson appears to have been one of the earlier post-Gödel workers to openly grapple with this issue and coined the phrase *explicitly definable* (Robinson 1949) to express a distinction with the sort of non-standard definition that will be necessary for the Incompleteness proof. There has since been a run in the literature on a number of related distinctions: representability, expressibility, complete representability, strong-representability, weak-representability, effective representability, assertability, capturability, correspondability, etc.

[12]   Out of dozens, we have yet to run across a review of GI that employs a Peano axiom set or alphabet as austere as Gödel's original. Always missing however are the higher-

case, we proceed to Gödel's now necessary method of PRF definition in P.

As a PRF "f" maps natural numbers to natural numbers, for every "n" there is an "m" that satisfies

(1) $f(n) = m$

a *relation* between the numbers in the general form of what is called a Primitive Recursive Relation (PRR), for which PRF, f, we denote F; i.e., F(n, m) ≡ (f(n) = m). We should like to have it then that for any PRR there will be some corresponding relation F in the syntax of P^{13}

$$\forall n \forall m. \, [\, F(n, m) \Leftrightarrow \mathsf{F}(\mathsf{n}, \mathsf{m})]$$

where the double-lined arrow "⇒" is the metalanguage symbol for material implication (i.e., the metalinguistic counterpart to "→" in P). Such an F corresponding to a recursively-defined f

$f(0) = g(0)$
$f(sx) = h(x, f(x))$

is true, if and only if there is *in P* some function β that satisfies

$$[\, \beta(0) = g(0) \cap (\, k < n \rightarrow \exists w \, \exists w' \, [\beta(sk) = w' \cap \beta(k) = w \,] \cap w' = h(k, w)) \cap$$
$$\beta(n) = m]$$

so that

$$f(n) = m$$
$$\Leftrightarrow$$

$$\exists \beta \, [\, \beta(0) = g(0) \cap (\, k < n \rightarrow \exists w \, \exists w' \, [\beta(k+1) = w' \cap \beta(k) = w \,] \cap H[k, w; w']) \cap \beta(n) = m].$$

"f" is of course one such β, though which unfortunately does not appear in P; were we to find in P such a β, or prove that there were one, then the

---

order variables that do appear in Gödel's signature. For discussion see reference (Shapiro 1991).

13   In Gödel's Theorem V there is a definability clause that makes this claim, which for the semantic version of Incompleteness discussed here, is the only feature of Theorem V relevant.

rhs of the above logical equivalence would be our desired relation F.[14] Such a β-function can in fact be derived in P for any PRR. The bracketed rhs above is a conjunction of simultaneous relations $\beta(k) = f_k$, in which each f_k satisfies, $f(k) = f_k$, having the form

(2) $[\cap_{i \le n} \beta(i) = f_i] \cap \beta(n) = m$

Given that multiplication and addition are definable in P, the congruence relation $x \equiv y (\bmod z)$ is also definable in P as[15]

$$\exists u \, [(u+x = y \,.\text{U.}\, y+u = x) \cap z \,|\, u]$$

I.e.,

$$x \equiv y (\bmod z) \Leftrightarrow \exists u \, [(u+x = y \,.\text{U.}\, y+u = x) \cap z \,|\, u]$$

In Modular Arithmetic there is a theorem that gives for any finite set of numbers $\{f_k\}$ the existence of a solution "x" to the simultaneous congruences

$$x \equiv f_i \bmod u_i \quad \forall \, i < n$$

provided the various u_i's are mutually prime, none evenly divisible into any other. This is the Chinese Remainder Theorem. Should we have it further then that, $f_i < u_i$, for all i, then an f_i is just the Euclidean "remainder" from the division of x by u_i.

$$f_i = x \bmod u_i \quad \text{(note equal sign)}$$

For β then we might try out the modular function

$$\beta(i) = d \bmod(u_i)$$

where the u_i's are all mutually prime, and "d" is the constant guaranteed to exist, so that from (2)

$$d \bmod (u_k) \equiv f_k , \qquad\qquad f_k < u_k$$

[14]   As H is a defining primitive of F, some G is a defining primitive of H, and for any beginning F this defining by relative primitives continues all the way to some combination of the functions, 0 and s, given in P as primitives. This is the kind of "inductive proof" of Definability (see term below) that Gödel envisages in his Theorem V, though he goes into no detail.

[15]   Where "z | u" is the proposition that z evenly divides u, and is defined in our "Capture" section below as, $\exists z < x. \, x = y \cdot z$, with, $\exists z(a) \equiv \neg \, \forall z(\neg \, a)$, and, $x < y \equiv \exists z \, (x + z = y)$

and f_k becomes the remainder of the division of d by u_k

$$f_k = d \bmod (u_k)$$

A mutually prime set $\{u_i\}$ adequate to our purpose would be

$$u_k - 1 \mid (k+1)\, c, \text{ with } c = \lceil \max(n; f_0, ..., f_n) +1 \rceil !$$

To summarize, then, given a recursively defined PRF "f"

$$f(0) = g(0)$$
$$f(sx) = h(x, f(x))$$

we have

$$f(n) = m$$
$$\Leftrightarrow$$

$\exists c\, \exists d\, \{\, [\, g(0) = d(\bmod u(0, c))\,] \cap [\, k < n \to \exists w\, \exists w'\, (w' = d(\bmod u(k, c)) \cap$
$w = d(\bmod u(k+1, c)) \cap H(k, w'; w))] \cap m = d(\bmod u(n, c))\}$

whereby the substitution

$$[x = y(\bmod z)] \Rightarrow \exists u\, [(u+x = y \,.\mathsf{U}.\, y+u = x) \cap z\,|\,u] \cap x < z$$

the rhs becomes the desired formula in P, $F(n, m)$, corresponding to the relation, $f(n) = m$. The β that appears here is known as the Gödel β-function.

How then does this work out for the particular case of the factorial function; what is its representation in P? With $g(0) = s0$ and $h(k, w') = w' \cdot sk$, we find immediately,

$F(n, m) \equiv$
$\exists c\, \exists d\, \{\, [\, d(\bmod u(0, c)) = s0\,] \cap [\, k < n \to \exists w\, \exists w'\, (\, w' = d(\bmod u(k, c)) \cap w$
$= d(\bmod u(k+1, c)) \cap w = w' \cdot sk)] \cap m = d(\bmod u(n, c))\}$

This form of F, recall, has been derived for an "f" that is recursively defined. For the PRF defined by composition of PRFs

$$f(x) = h(g(x))$$

the corresponding F is given simply as[16]

---

[16]   I.e., $h(g(n)) = m \Leftrightarrow \exists z\, [\, g(n)=z \cap h(z)=m]$

$$\exists z[\ G(n,\ z) \cap H(z,\ m)]$$

We use the term "definition" above to refer to the definition, expression and representation of a PRR in the language of P. We now define Definition by the "PRR $\Rightarrow$ P" correspondence, such that for all PRRs, R,

$$\exists a \in P.\ R \Rightarrow a$$

to be read as,

R is Definable in P

which is close to Gödel's terminology, though the usage in (Gödel 1962) is not always consistent.[17]

In the coming subsections we consider two additional relation-correspondences with which Definition must not be confused: in the next subsection, a meta-P $\Leftrightarrow$ meta-P' correspondence between statements related by the G-transformation,

$$a \Leftrightarrow b$$

which we will read as

a and b Correspond

and in the subsection that follows, a meta-P' $\Leftrightarrow$ P correspondence

$$a \Leftrightarrow b$$

which we read as[18]

a is Captured by b

From now we use the term "PRR" to refer both to a PRR and its definition in P.

Building PRFs and PRRs

As PRRs will later be derived to correspond with meta-P' propositions, one will need methods by which new PRRs and PRFs are constructed from

---

[17]   As F *Defines* F ($\equiv$ f(n) = m) in P, one will sometimes hear that it *Represents* f, as the abstract mapping pairs of f gives its extension.

[18]   This $a \Leftrightarrow b$ reads in (Gödel 1962), for the most part, as "a is *expressed* by b".

existing ones. Given the form, $f(n) - m = 0$, we may say that a PRR can always be given as a rooting of some PRF,[19] and that, generally, corresponding to every PRR is a characteristic function (CF), φ, that satisfies

$$R(n, m) \Leftrightarrow \varphi(n, m) = 0$$

where ϕ is a PRF. Using the rooting of PRFs now to identify PRRs derives five methods for PRR and PRF construction.

i)   The disjunction of PRRs, $R \equiv R_1 \cup R_2$, is itself a PRR, for its CF is immediately given as $\varphi \equiv \varphi_1 \cdot \varphi_2$ i.e.,

$$[\, (\, R_1 \Leftrightarrow \varphi_1 = 0\,) \cup (R_2 \Leftrightarrow \varphi_2 = 0)\,]\,\Rightarrow\,[\,R_1 \cup R_2 \Leftrightarrow \varphi_1 \cdot \varphi_2 = 0\,]$$

which holds equally of course for any finite number of disjoined PRRs.

ii)   By means of the step function,

$$\text{step(n)} = \begin{cases} 0, & n = 0 \\ 1, & n \neq 0 \end{cases}$$

recursively given as

$$\text{step}(0) = s0$$
$$\text{step}(sx) = 0$$

it is clear that the negation of a PRR, $\neg R$, is itself (by composition) a PRR

$$[\,R \Leftrightarrow \varphi = 0\,]\,\Rightarrow\,[\,\neg R \Leftrightarrow \text{step}(\varphi) = 0\,]$$

iii) As any logical operation between propositions, including PRRs, is reducible to negations and disjunctions, any logical combination of PRRs, then, is also a PRR; we find its CF via i and ii above.

iv)   Bounded quantifications of a PRR, then, which may always be given as the logical conjunctions or disjunctions

$$\forall x\,(\,x \leq n \cap R(x, m)) \equiv R(1, m) \cap R(2, m) \cap .... \cap R(n, m)$$
$$\exists x\,(\,x \leq n \cap R(x, m)) \equiv R(1, m) \cup R(2, m) \cup .... \cup R(n, m)$$

are themselves PRRs, and in particular

$$\forall x\,(\,x \leq n \cap R(x, m)) \Leftrightarrow \varphi_u(n, m) = 0$$
$$\exists x\,(\,x \leq n \cap R(x, m)) \Leftrightarrow \varphi_E(n, m) = 0$$

where

---

[19]   Like the addition operator, the difference operator $\chi(x, y) = x - y$ is also a PRF, so that by composition $\chi(\,f(n), m)$ is also a PRF.

$$\varphi_u(n, m) \equiv \Sigma_{i=0}^n\, \varphi(i, m) = \varphi(1, m) + \varphi(2, m) + \varphi(3, m) + \ldots + \varphi(n, m)$$
$$\varphi_E(n, m) \equiv \cap_{i=0}^n\, \varphi(i, m) \equiv \varphi(1, m) \cdot \varphi(2, m) \cdot \varphi(3, m) \cdot \ldots \cdot \varphi(n, m)$$

are the respective bounded universal and bounded existential CFs.

v)   Finally, for a given PRR, R, we define a particular PRF, (min: x <n ∩ R(x)), by its mapping to the smallest x < n that satisfies R(x), or should there be no such x < n, it maps simply to n.

$$(\text{min: } x < n \cap R(x)) = \begin{cases} m, & R(m) \cap [\neg\, \exists r\, (r < m \cap R(r)] \\ n, & \neg\, \exists m[\, m < n \cap R(m)] \end{cases}$$

From the above expressions we can see that the desired function may be constructed (for R ⇔ φ = 0) as a composition of bounded Universal and bounded Existential CFs

(min: x <n ∩ R(x)) ≡ $\Sigma_{i=0}^n$ step(step($\varphi_E(i, m)$)) = step(step($\varphi_E(1, m)$)) + step(step($\varphi_E(2, m)$))
+step(step($\varphi_E(3, m)$)) + ... + step(step($\varphi_E(n, m)$))

where step(step(x)) = 0, for x = 0, and step(step(x)) = 1, otherwise, so that (min: x <n ∩R(x)) as a composition of PRF compound compositions is itself a PRF.

Correspondence

meta-P' ⇔ meta-P

The formal system P is defined by the alphabet

$$\{0,\, s,\, n,\, m\, f_2,\, f_3,\, f_4,\ldots,\, (\, ,\,),\, \cup,\, \forall,\, \neg\, \}$$

together with formation rules and axioms stated above. A transformation of individual letters alone interprets the same model (e.g., the natural numbers) only in a different symbolism. E.g., while in the alphabet of P, the assertion "3+5=8" is rendered

$$sss(sssss0) = sssssss0,$$

in an alphabet

$$\{0,\, s,\, n,\, m\, f_2,\, f_3,\, f_4,\ldots,\, p\, ,\, q\, ,\, \cup,\, \forall,\, \neg\, \}$$

it is rendered

$$ssspsssssOq = sssssssssO.$$

Even such a minor alteration defines a distinct formal system, some $\tilde{P}$, this particular one being isomorphic to P via the simple syntax mapping,

$$I : \{(,)\} \longrightarrow \{p, q\}$$

which yields, again, the very same theory of natural numbers, though expressed in a different script. Out of an infinity of such possible symbolisms for arithmetic, our system P and Gödel's special isomorphism G (which we shortly consider) selects a unique pair, P and P' - one, under G, the isomorphic image of the other; likewise, their respective meta-languages, meta-P and meta-P'.[20]

By the fundamental theorem of arithmetic, every natural number has a unique factorization into orders of prime numbers, as

$$1200 = 2^4 \cdot 3^1 \cdot 5^2$$

the example offered by Wikipedia (Contributors, Fundamental Theorem of Arithmetic 2021) in *standard form*, the base primes ordered by magnitude, left to right. While the factorizations of 1200 are endless, there is no other in primes; the fundamental theorem is also known as the unique factorization theorem, which over the years has seen many applications. For a secret message, you might pass on the numeral "1200". Assuming your accomplice has a deciphering key: 1 ↔ a, 3 ↔ b, 5 ↔ e, 6 ↔ g, 4 ↔ m, 2 ↔ n, 7 ↔ r, and a working knowledge of the fundamental theorem of arithmetic, they will understand you to mean "man" and presumably know what to do. Given the key, such coding of any sequence of the alphabet {a, b, e, g, m, n, r} is unambiguous.

On the other hand, not all numbers can be deciphered with our limited key; not all numbers are *code numbers*, since the largest decipherable prime exponent is 7—for example, the number whose prime factorization is $2^{2.4666111e+21} \cdot 3^4 \cdot 5^1 \cdot 7^2$. This limitation opens however the possibility for deeper coding, not only of the individual letter, but of letter sequences. Say you send the number $2^{2.4666111e+21} \cdot 3^4 \cdot 5^1 \cdot 7^2$ with the further mutual understanding that any prime exponent >7 itself codes a word. The message will then be decoded

---

[20]   We will see the importance in our coming "Indirect Self-Reference" section whether P and meta-P are also mutual isomorphic images, under G or otherwise.

$2^{2.4666111e+21} \cdot 3^4 \cdot 5^1 \cdot 7^2 \rightarrow 2.4666111e+21, 2^4 \cdot 3^1 \cdot 5^2 =$
$$2^3 \cdot 3^5 \cdot 5^6 \cdot 7^6 \cdot 11^1 \cdot 13^7, 2^4 \cdot 3^1 \cdot 5^2 \rightarrow \text{beggar man}$$

The same embedding may be applied from words to sentences, the next syntactic level up, and so on.

This is Gödel's scheme for the G-transformation of P, of which are two components, g_1 and g_2:

g_1 is the *alphabet number-coding*

g_1: $\{0, s, \neg, \cup, \forall, (,), n, m\ f_2, f_3, f_4,...\} \rightarrow \{1, 3, 5, 7, 9, 11, 13, 17^1, 19^1, 23^2, 29^3, 31^4, ...\}$

whereby a Latin numeral in Comic Sans MS typesetting, n, we mean the P-numeral "*sss...s0*" in which "*s*" appears n-times. Furthermore, variables of type "i" are mapped to a prime > 13 raised to the ith power.

g_2 is a *number-sequence number-coding* that takes an n-length number-sequence resulting, for example, from a g_1 transformation of an n-length formula to a product of primes of increasing magnitude, left to right, each raised to the power of the number in its ordered place, so that

g_2: $11, 3, 3, 1, 13 \rightarrow 2^{11} \cdot 3^3 \cdot 5^3 \cdot 7^1 \cdot 11^{13} =$

38378687463734344789018

The G-transformation of P then proceeds in three targeted stages that may be visualized by application to the above formalization scheme diagram:

i. g_1 is applied to all letters in P columns: alphabet, signs, formulas, axioms, and proofs

ii. g_2 is then applied to the resulting number sequences in P columns: signs, formulas, axioms, and proofs; and finally

iii. g_2 is re-applied to the resulting number sequences in P column proofs.

Thus, for $a_1 \equiv sss(sssss0) = sssssss0$, $a_2 \equiv sssss(ssss0) = ssssss0$, and $a \equiv a_1, a_2$

$$G[0] = g_1[0] = 2$$
$$G[a_1] = g_2[g_1[a_1]]$$
$$G[a] = g_2[g_2[g_1[a_1]], g_2[g_1[a_2]]$$

The resulting language is our P', and just as there is a meta-P language, there is a corresponding meta-P' language.

P'

alphabet	signs	formulas	axioms	proofs
1	:	:	:	:
3	17	:	:	:
17	:	:	$G[\forall n\neg(sn=0)]$	:
19	:	$G[f_3(f_2(n))]$	:	------------
$\{G[f_i]\}$	$2 \cdot 3^3 \cdot 5^3 \cdot 7^3 \cdot 11^1$	:	:	:
13	:	:	:	:
11	:	$G[f_2(0)\cup f_2(sm)]$	:	:
7	37^5	:		------------
9	:	:		$G[\forall n\neg(sn=0)]$
5	:	:		:
	:			:
				$G[sss(sssss0) = sssssss0]$

				:
				:

Once the G-transformation is outlined, Gödel opens:

> Now let $R(\{a_i\})$ be a given class or relation between basic "signs" or sequences of them. We will associate that with the class (relation) $R'(\{x_i\})$ that holds between $\{x_i\}$ if and only if there are $\{a_i\}$ such that for all j, we have $x_j = G[a_j]$ and the $R(\{a_i\})$ holds.

Thus, given, $x = G[a]$, the association

$$R(a) \Leftrightarrow R'(x)$$

is purely analytic, meaning nothing more than

$$R'(x) \equiv R(G^{-1}[x])$$

an association between meta-P statements that recognize the G-transformation.[21]

---

[21]   The paragraph is often mistaken as a statement of a kind of Capture correspondence that associates with every $R(a)$ a guaranteed PRR, $R'(x)$, e.g., ref. (FitzPatrick 1966).

For example[22]

$\mathfrak{a}$ is a variable in P $\iff$ $G^{-1}[x]$ is a variable in P

or in the meta-P/meta-P′ correspondence

$\mathfrak{a}$ is a variable in P $\iff$ x is a variable in P′

which we represent by the notation

$\mathfrak{a}$ is a variable $\iff$ x is a VARIABLE

an association between meta-P/P′ statements given by the G-transformation. This material equivalence exemplifies the truth-preserving G-isomorphism between meta-P and meta-P′ statements. As we shall see in the next subsection, Gödel also manages to establish material equivalences between *certain* meta-P′ statements and P-statements—*arithmetical* statements. It is sometimes convenient to imagine the meta-P statements that satisfy these equivalences taken together

$$\text{meta-P} \iff \text{meta-P′} \iff \text{P}$$

as *translations* or alternate *interpretations* of the corresponding P-statement's arithmetical syntax, which indeed is the obvious sense of the phrase "arithmetization of the metalanguage". They are not.[23] What is referred to as Gödel's arithmetization of the metalanguage of P is more correctly only a *numeralization*, rendered not in arithmetic, but in ordinary language. We are far from the first to point this out.

In the 1968 *What the Gödel Formula Says* (Lacey and Joseph 1968, 77-83), we find for example:

> ... the account cannot be sustained, for it is not possible to consider some of the axioms of P (for example, the Peano axioms) as having interpretations in M. It follows, therefore, that the only relevant, available interpreted system, P', is in the

---

[22]   Gödel renders the correspondence
$(x = G[\mathfrak{a}])$ $\Rightarrow$ ('$\mathfrak{a}$ is a variable' $\iff$ '$G^{-1}[x]$ is a variable')
as
x is a *variable*
and others in some like shorthand.

[23]   For one thing, whereas the sense of Gödel's derived 46 relations given in the "Capture" subsection below would be clear, there would be no sense to such an alternate meta-P translation of '3+5=8'. And, of course, not all numbers as represented in P are G-numbers.

language of A, since a formal system cannot be considered as an interpreted system unless all of its axioms and theorems receive some interpretation.

And from early reviewer Braithwaite (Gödel 1962):

> Gödel's 'arithmetization' of metamathematical concepts (as also my 'modified arithmetization') is in fact effected by correlating to each string x another string which is a numeral: there is no need to pass from a string x to this numeral by the indirect route of first moving to the Gödel number (or G-number) of x and then passing from this number to the numeral which expresses it in the calculus P.

It is Gödel's own above quoted paragraph that is often mistaken as a statement of such a meta-P/P *translation*. If only there were such a translation! Then, with only the metalogical sense of nonprovability, we could go straightaway to the existence of a nonprovability P-predicate and in one additional stroke to semantic Incompleteness—without deriving a single of the 45 PRRs that comprise the bulk of Gödel's actual work.

Meta-P propositions are not supplanted by arithmetical propositions, rather they are *only the subjects* of meta-P propositions, elements of P, that are supplanted by numerals, elements of P'. The distinction is irrelevant to the establishment of Incompleteness itself, and indeed the mistaken conception of arithmetization, which has largely prevailed over the years, might well lend some expository advantage. As will become apparent in the "Indirect Self-Reference" section however, it can indeed be relevant to wider implications of GI.

Capture

P ⇔ meta-P'

Now with P' and relations between PRRs and PRFs, Gödel proceeds in rapid succession through 45 PRF and PRR definitions of increasing complexity, one built upon another. The first simple several are ancillary to construction of more complex PRRs that Capture meta-P' propositions, ending with a concatenation operator "∘", number eight on the list. The next are "syntax" PRs needed for the construction of a self-referring apparatus, ending with a substitution PRF, $subst(x, v, y)$, #31. The final project, "deduction" PRR's, build to the construction of the provability predicate, $prov(x)$, a #46th non-PRR. Given that PRR's are Definable in P, we shall use the same name for a PRR and its P-Definition, using ordinary and Comic Sans MS typescript interchangeably, as convenient. We consider here a few key representatives from the 46; the full list may be found at reference (Gödel 2000).

Ancillary PRRs

The first set of definitions are also of a varied sort. Like the first, some are purely arithmetical:

1.)
$$y|x \equiv \exists z < x.\ x = y \cdot z$$

which meaning is given by the standard interpretation, "y divides evenly into x", whether or not x or y are G-numbers.[24] Along with others, these are the bricks and mortar of later, more interesting, definitions. Others built out of these, while having ordinary arithmetic interpretations, are also in correspondence with elements of P, provided they are G-numbers, thus objects that might appear in meta-P' statements. For example, given that the *purely arithmetical* function that $prFactor(n, x)$ maps $\{n, x\}$ to the nth prime factor of x in standard form (third on the list), the function

6.)
$$item(n, x) \equiv min:\ y \leq x.[(prFactor(n, x)^y \mid x) \cap \neg$$
$$(prFactor(n, x)^{y+1} \mid x)\ ]$$

maps $\{n, x\}$ to the exponent of the nth prime factor of x, again, whether or not x is a G-number. If x is a G-number, then $item(n, x)$ is the G-number of the nth letter of $G^{-1}[x]$.

   To avoid redundancy in what follows, we now define the G-numbers x, y, z, v', by their P counterparts:

$$x = G[a],\ y = G[b],\ z = G[c],\ v' = G[v]$$

Yet other ancillaries are proposition functions on the natural numbers that also Capture meta-P' proposition functions. Given a proposition

---

[24]   Recall that in our notation '$\equiv$' is the symbol for formal definition and '$\Leftrightarrow$' for material equivalence. While symbolism for the 46 in (Gödel 1962) and (Gödel 2000) differ, their meanings are the same. The meanings here differ; rather than as shorthand for 'y divides evenly into x', $y|x$ is taken here as shorthand for the syntax $\exists z < x.\ x = y \cdot z$.

function isPrime(z) (second on the list) that corresponds to the arithmetical property of being a prime number, the proposition function defined as

11.) $vtype(n, x) \equiv \exists\ 13 < z \leq x.\ [\ isPrime(z) \cap x = z^n) \cap n \neq 0\ ]$

Captures the meta-P' proposition that x is a variable of type n, i.e.,

$$vtype(n, x) \Leftrightarrow x \text{ is a VARIABLE of type } n$$

so that

12.) $isVar(x) \equiv \exists n < x.\ vtype(n, x)$

Captures the meta-P' proposition that x is a VARIABLE

$$isVar(x) \Leftrightarrow x \text{ is a VARIABLE}$$

which itself corresponds to the meta-P proposition that a is a variable

$$x \text{ is a VARIABLE} \Leftrightarrow a \text{ is a variable}$$

What does isVar(x) *say*?

This is our first propositional Capture: let us clarify just what that means. isVar(x) is the formal definition that designates the unmanageable PRR expression

$$\exists n \leq x.\{(\exists 13 < z \leq x.[\ \neg(\exists r \leq z.(r \neq 1 \cap r \neq z \cap [\exists q \leq z.\ z = r \cdot q]\ )\ )\cap$$
$$(r > 1)] \cap x = z^n) \cap n \neq 0$$

which reads, "There is a natural number less than or equal to x, such that...", or less formally, "x is some natural number raised to the power of a prime greater than 13". Its meaning, its intention, then, as an *arithmetical statement* is clear.

On the other hand, the statement is specially constructed. The P-system unary numeral representation of the x that satisfies isVar(x) is a variable in the language P', so that the truth or falsity of the arithmetic statement isVar(x) is directly correlated to and so may be *inferred* from the truth or falsity of the meta-P' statement, "x is a variable in P'" i.e., the statements are *extensionally* equivalent (in the conventional Fregean sense (Fitting 2020)), and that is what the Capture statement

$$isVar(x) \Leftrightarrow x \text{ is a VARIABLE}$$

means, and the reason for the mnemonic denotation, *isVar*.

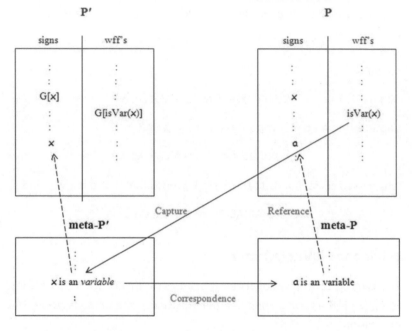

Had Gödel opened the Incompleteness derivation with

All fruit have seeds & Apple is a fruit ⇒ Apples have seeds

one might have wondered at the relevance, but no-one would have mis-understood the sense nor doubted the truth of the conditional. Further-more, the sense of the consequent is *contained* in that of the antecedent (which may be seen explicitly by expanding the universal quantifier "all" as a conjoined series of statements ranging over fruit), so that, "All fruit have seeds & apple is a fruit" already says "Apples have seeds".

On the other hand, had Gödel set out with

$$\exists n \leq x.\{(\exists 13 < z \leq x.[\neg(\exists r \leq z.(r{\neq}1 \cap r{\neq}z \cap [\exists q \leq z. \ z = r{\cdot}q]))\cap$$
$$(r >1)]\cap x{=}z^n)\cap n{\neq}0 \ \Leftrightarrow \ x \text{ is a VARIABLE}$$

one would hardly know what to make of it, and much less its truth-value. As it happens, by the time the reader has reached the ninth page where the above identity is actually given, he or she has already seen a good deal and understands very well what the *premises* material to the conditional are: the formal system P and the syntax mapping G, which are implicitly read into it (e.g., in the manner of Enthymemes (Anderson and Belnap 1961)):

$$\{G, P\} \vdash [\text{ isVar}(x) \Leftrightarrow x \text{ is a VARIABLE}]$$

where, again, the double-lined arrow signifies synthetic, material implication, whose *material* is precisely {G, P}. To the conditional, apple is a fruit $\Rightarrow$ apples have seeds, is likewise material the premise that fruit have seeds, so that

$$\{\text{All fruit have seeds}\} \vdash [\text{apple is a fruit} \Rightarrow \text{apples have seeds}]$$

but we certainly do not then have, "apple is a fruit" *says* that "apples have seeds". In both cases, the senses, the intensions, of the consequents are entirely distinct from those of the antecedents, and one does not *say* what the other says. isVar(x) does not say "x is a variable in P'". The Capture correspondence (like Correspondence) is a material and not an analytical containment relation (Correia 2004).

Other ancillaries still are operations on the natural numbers, n-ary mappings, that correspond to syntactic mappings in P'. For example, given a function $\text{length}(x)$ (seventh on the list) that maps x to the number of terms in its prime factorization (and so to the number of letters in a) the operator "$\circ$" defined by

8.) $$x \circ y \equiv \text{ min } z \leq \text{nthPrime}(\text{length}(x) + \text{length } (y))^{x+y}$$
$$.(\forall n \leq \text{length}(x). \text{ item}(n, z) = \text{item}(n, x)) \cap$$
$$(\forall 0 < n \leq \text{length}(y).\text{item}(n + \text{length}(x), z) = \text{item}(n, y))$$

maps {x, y} to a number whose first standard-form exponents are those of x, and last are those of y. $x \circ y$ then is the G-number of ab.

$$x \circ y = G[ab]$$

Syntax PRRs

Gödel next develops components for a kind of self-referencing that involve the predication of a G-number by its P predicate—a predication of the form, $a(x)$. In this there are two key functional components. One is the PRF, $number(x)$, recursively given as

17.)
$$number(0) = 2$$
$$number(Sn) = 2^3 \circ number(n).$$

Thus $number(x)$ is the G-number of x; i.e., $number(x) = G[x]$, so that $number$ maps P' numbers as the G-transformation maps P-numerals.

The other key component is a mapping, $subst$, that provides a means to identify P' propositions that are derived by free-variable substitution. The corresponding meta-P $Subst[a](v, b)$ denotes the P-expression a, whose free-variable v is substituted by b. Its own two basic functional components are

i)   a function $freePlace(k, v', x)$ (28 on the list) that maps to the place number, counted from *right to left*, of the $k+1^{st}$ appearance in a of the free-variable v.

E.g., for
$$a \equiv a_1 \, a_2 \, a_3 \, ... \, v \, ... \, a_m,$$
where a_i is the ith letter in a, as counted *left to right*, and v is the nth letter.

$$freePlace(0, v', x) = m-n+1;$$

and

$$freePlace(q, v', x) = 0 \text{ for } q \neq 0$$

so that the function

29.) $nFreePlaces(v', x) \equiv \min: n \leq length(x).free\text{-}$
$Place(n, v', x) = 0$

maps to the number of places in a where the variable v is free.

and

ii) a function $insert(x, n, y)$ (#27 on list) that maps to the G-number of an altered a in which its nth letter has been substituted by b. I.e.,

$$insert(x, n, y) = G[a_1\ a_2\ a_3 \ .... \ b \ .... \ a_m].$$

Assuming we have these two functions then, the function $subst'(n, x, v', y)$ recursively defined as

30.)
$$subst'(0, x, v', y) = x$$
$$subst'(k+1, x, v', y) = insert(subst'(k, x, v, y), freePlace(k, v, x), y)$$

maps to the G-number of an altered a in which the first 'n' occurrences of v, right to left, have been substituted by b.

Thus,

31.) $$subst(x, v', y) \equiv subst'(nFreePlaces(v', x), x, v', y)$$

maps the G-number of $a(v)$ to the G-number of $a(b)$, i.e. [25]

$$subst(x, v', y) = G[a(b)]$$

Diagonal Composition

$P_{fp} \Leftrightarrow P$

We define the composition PRF

$$diag(x) \equiv subst(x, v', number(x))$$

that maps x to the G-number of a evaluated at x

$$diag(x) = G[a(x)].$$

It maps to the G-number diagonal of a *P-function* by *P-function G-number* matrix

---

[25]   To more closely correspond to the informal definition given for the meta-P function Subst, one might amend the above definiens with the condition " .... ∩ ∃n [stype(n, y)∩stype(n, v)]", to insure that v' and y are of the same Type.

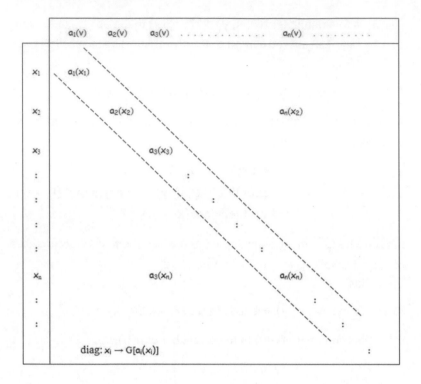

and we say that $\text{diag}(G[a])$ is the *diagonalization* of a. Let us consider the effect of PRR composition with diagonalization on Capture and Correspondence relations using the example of the $\text{isVar}(x)$ proposition-function from our last section.

P		meta-P′		meta-P
$\text{isVar}(x)$	⟺	x is VARIABLE	⟺	a is a variable.
$\text{isVar}(\text{diag}(x))$	⟺	$\text{diag}(x)$ is a VARIABLE	⟺	$a(x)$ is a variable.

where recall that x is the G-number of a. We define the PRR composition $\text{Vs}(x) \equiv \text{isVar}(\text{diag}(x))$, so that

$$\text{Vs}(x) \qquad\qquad ⟺ \qquad\qquad a(x) \text{ is a variable.}$$

and

$$\text{Vs}(G[\text{Vs}]) \qquad\qquad ⟺ \qquad\qquad \text{Vs}(G[\text{Vs}]) \text{ is a variable.}$$

By this material equivalence, $Vs(G[Vs])$ is sometimes said to refer to itself.

What does $isVar(diag(x))$ say?

The sense of $isVar(diag(k))$ follows directly from the strict arithmetical sense of isVar(x) discussed above in the sub-section "What does $isVar(x)$ say?"; it says nothing more. In particular, while it may be *deduced* from $isVar(diag(k))$ that $isVar(diag(k))$ is a variable (given the material P and G), $isVar(diag(k))$ does not *say of itself* that it is a variable.

Fixed-Point

P ⇔ meta-P' ⇔ meta-P → P$_{fp}$ ⇔ P

There is nothing particular about $isVar(x)$ in the above self-ref mechanism, and we may choose any PRR that Captures some meta-P' property to obtain a corresponding result. Gödel derives it only for the ¬prov(x) PRR of the next subsection, which Captures the provability property of interest, and via the specific deduction mechanism summarized in figure (B) of sub-section "Konigsberg, 1930", below. We briefly review the general case here for the sake of the discussion in the coming "II. Indirect Self-Reference" section.

For an arbitrary P-predicate function a, define $adiag(v)$ as the composition $a(diag(v))$, and x_{ad} as its G-number

$$adiag(v) \equiv a(diag(v))$$
$$x_{ad} \equiv G[adiag(v)]$$

From $diag(x) = G[a(x)]$,

$$adiag(x_{ad})) \equiv adiag(G[adiag(v)])) \Leftrightarrow a(G[adiag(G[adiag(v)])]) \equiv$$
$$a(G[adiag(x_{ad})]).$$

the correspondence

$$a(G[adiag(x_{ad})]) \Leftrightarrow adiag(x_{ad})$$

holds for any property a. We denote this correspondence Diagonal Composition (Diag), known elsewhere as the diagonal lemma (Simmons 1993). Its form is reminiscent of the Fixed-Point Theorem of analysis which states that under certain conditions a numerical function "f" will have at least one "α" in its domain that satisfies

$$f(α) = α$$

Such an α is said to be a fixed point of f. Using this language, we may read from

$$adiag(x_{ad}) \Leftrightarrow a(G[adiag(x_{ad})]) \Leftrightarrow adiag(x_{ad}) \text{ is an } a$$

that for every propositional function a in P there is another proposition in P, $adiag(x_{ad})$, its *fixed point*, in material identity with the assertion that it satisfies a. We diagram the relevant structure.

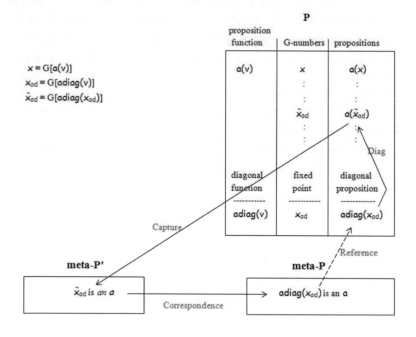

Incompleteness

To find the Gödel sentence it remains only to derive a PRR that Captures the meta-P' property of being un-provable in P'; g would then be the fixed point of its composition with the G-transformation (yielding to the meta-language), so that

$$g \Leftrightarrow g \text{ is not provable}$$

which if true, g would be true but not provable, and if false, g would be false and provable, which, assuming P to be sound, is not an option.

For a "provability" predicate, we must finally consider the business of axioms and formal deduction in P. Aside from briefly mentioning three of the nine Peano axioms (for illustration in the section "Formal System P") we have avoided this subject that in (Gödel 1962) comes early on in the form of axiom schema.

Gödel selects a formulation just adequate to the task. Rather than the full nine Peano axioms, only the first three appear; the others are accounted for by logical axiom deduction in combination with the three. This non-logical axiom selection is *perhaps* the most economical possible (cf. footnote (12)).[26] We have avoided the subject because the particulars of the deduction apparatus are not significant, only that it be sufficient to derive the predicate calculus arithmetic we have so far been using—the distributive property of natural numbers, the associative property, etc.— and specifically for the definition of PRRs. We proceed then, accepting Gödel's selection.

Deduction PRRs

Assuming that we have

i)   a PRR $isAxiom(x)$ (42 on the list) that Captures the meta-P' proposition

$$x \text{ is an AXIOM}$$

and

---

[26]   One of the recurring objections to the derivation has been Gödel's the use of PM's Reducibility Axiom, which is not essential; one may choose any number of axiom sets that do not include the reducibility axiom, as most do not.

ii) a PRR immConseq(x, y, z) (43 on the list) that Captures the meta-P′ proposition

$$x \text{ is DEDUCIBLE}^{27} \text{ from } y \text{ and } z$$

corresponding to the meta-P proposition

$$a \text{ is deducible from } b \text{ and } c$$

then, isProofSequence defined by

44.)   isProofSequence$(x) \equiv [\forall 0 < n < \text{length}(x). \text{isAxiom(item}(n, x))$
$\cup \exists 0 < p,q < n. \text{immConseq(item}(n, x), \text{item}(p, x), \text{item}(q, x))]$
$\cap \text{length}(x) > 0$

Captures the meta-P′ proposition

The ITEMS of x are DEDUCTIONS from ANTECEDENT ITEMS or AXIOMS.

More precisely, defining, $x = G[A]$, isProofSequence(x) Captures the meta-P′ proposition:

$$x \text{ is a PROOF SEQUENCE.}$$

Building upon these, Gödel's 45th proposition-function

45.) proves$(y, x) \equiv$ isProofFigure$(y) \cap \text{item(length}(y), y) = x$

Captures the meta-P′ statement

$$x \text{ appears as the FINAL ITEM in PROOF-SEQUENCE } y,$$

the property of being the final formula in a proof-sequence, and his final proposition-function, and

46.)   prov$(x) \equiv \exists y.$ proves(y,x)

Captures the property of being provable, so that

$$\neg \text{ prov}(x) \Leftrightarrow x \text{ is NOT PROVABLE}$$

from which the statement g is immediately derived

---

[27]   Via Modus Ponens or Universal Introduction.

$$g \equiv \neg \ prov(diag(G[\neg \ prov(diag(v)) \ ]))$$
$$\Leftrightarrow \neg \ prov(G[g])$$

with its part as the fixed point of ¬**prov** given by GI

$$g \Leftrightarrow g \text{ is not provable}$$

From the most basic Peano axioms of arithmetic expressed in the formal system of sub-section "PM's Formal System P" above, we have derived a non-provability fixed-point relation, (the engine that powers Gödelian Incompleteness) whose indirect self-reflection immediately invites a speculative reading.

II. Indirect Self-Reference

Verification of the Incompleteness claim has involved an informal inference from the metalinguistic result

$$y \leftrightarrow g \text{ is not provable}$$

that summarizes the sequence of correspondences

$$P \Leftrightarrow \text{meta-P}' \Leftrightarrow \text{meta-P} \rightarrow P_{\text{fp}} \Leftrightarrow P$$

we denote Capture, Correspondence, and Diagonal Composition.

Rautenber's *Concise Intro to Mathematical Logic* (Rautenburg 2006, xvii) aptly characterizes the derivation:

> It is an intellectual adventure of holistic beauty to see wisdom from number theory known for ages, such as the Chinese remainder theorem, simple properties of prime numbers, and Euclid's characterization of coprimeness, unexpectedly assuming pivotal positions within the architecture of Gödel's proofs.

Anyone who has read Gödel's original work and presumes some understanding of it is tempted to write a GI paper. In the course of their study, they will encounter a range of diverging views and rationales—some good, some bad, some wrong for the right reasons, others right for the wrong reasons. Furthermore, there are many ways for an actual derivation of GI to go wrong, many opportunities for misinterpretation, and it is instructive to sort through a few.

"One is naturally tempted to get around it [Incompleteness]—which is indeed the only way to understand it" (Girard 2011, 15). In the remaining space of this sub-section we look at set of such views that we meet online. The thread we've chosen quite at random starts out on a sci-logic forum site (Jones 2007) and follows the comments of a Ms. LauLuna and Messrs. Hoge, Smith and Jones.

A common error that leads to misinterpretation of GI is the mishandling of the diag function, in particular its evaluation at $k = G[a(\text{diag}(v)]$:

i) $$\text{diag}(k) = G[a(\text{diag}(k))]$$

There is sometimes confusion whether diag(k) appears on the rhs as a single numeral or as a numeral expression, the difference being that although

$$2 \cdot 3 = 6$$

and that for any arithmetic predicate a,

$$a(2 \cdot 3) \Leftrightarrow a(6),$$

yet

$$G[2 \cdot 3] \neq G[6]$$

and

$$G[a(2 \cdot 3)] \neq G[a(6)].$$

To clarify the distinction we here underline single numerals where there is risk of ambiguity. For the diag function we have then:

ii) $\underline{diag(k)} = diag(\underline{k}) = G[a(diag(\underline{k}))]$

(where the expression diag($\underline{k}$) maps to the number $\underline{diag(k)}$) and, taking the two outermost terms, more explicitly[28]

$$sssss....s0 = G[\ \exists n(A(n) \cap Diag(sss...s0, n)]$$

where the successor operator "s" appears k-times on the rhs, and G[$\exists n(A(n) \cap Diag(sss...s0, n)]$-times on the lhs.

By contrast, i) is commonly misread as either

$$\underline{diag(k)} = G[a(\underline{diag(k)}\ )]$$

or

$$diag(\underline{k}) = G[a(diag(\underline{k}))] = G[a(G[a(diag(\underline{k}))])] = \ ....$$

that in effect takes the string "G[a]" to denote a numerical mapping (rather than a syntax-mapping), and we term the deviations "G-numerical(1)" and "G-numerical(2)", respectively, both of which give spurious infinities.

---

[28]   cf. notation in *Definition* subsection above.

The forum opens with Jones in one corner holding forth on the "monadic" nature of Gödelian self-reference, that its g, like an illicit Liar sentence (Smith 2013, 4), invalidates the Incompleteness claim. Smith, in the opposite corner, in due course counters:[29]

> Again wrong... The relevant unprovable sentence g that demonstrates P's Incompleteness is a sentence of elementary arithmetic, technically a Π_1 sentence, of a kind that even a Hilbertian finitist would recognize as meaningful. That sentence is about numbers, nothing else. This Gödel sentence g does *not* refer to itself, it's about numbers....

This carries on through several additional postings. Meanwhile on a linked sci-logic site (Hoge, New 2010), a Mr. Hoge introduces his "New Essay on the Gödel's Incompleteness Theorem", where GI proceeds via his newly coined "supernatural numbers", whose existence derives from the diag function relation:

$$diag(k) = G[a\ (diag(k))] = G[a(G[a(diag(k))])] = G[a(G[a(G[a(diag(k))])])] = \ldots\ldots$$

Ms. LauLuna, entering the discussion late, advises, "No":

> That's wrong. $a(diag(k))$ is about $diag(k)$, that is, the Gödel number of the result of substituting k for x in the formula whose Gödel number is k-the formula $a(diag(x))$-, a result which is precisely $a(diag(k))$... [so that] The circle is, so to say, completed.

We point out here that the statements $a(diag(\underline{k}))$ and $a(\underline{diag(k)})$ are logically equivalent, as are their meta-logical counterparts:

$$a(diag(\underline{k})) \Leftrightarrow G^{-1}[k] \text{ is an } adiag$$
$$a(\underline{diag(k)}) \Leftrightarrow G^{-1}[diag(k)] \text{ is an } a.$$

There is nothing decided here by choosing what $a(diag(k))$ is "about", and there is also the danger that in other contexts a sanctioned choice will lead to trouble. For example, simple counting demonstrates the untenability of its direct application to the diag relation i):

---

[29]   The same Peter Smith now thought to be a leading GI expositor (see for example subsections "Königsberg, 1930" and "Lisbon & Pisa, 2014" below). Peter Smith in the trenches.

$$\underline{diag(k)} = G[a(\underline{diag(k)})].$$

As the lhs numeral $\underline{diag(k)}$ is composed of "$\underline{diag(k)}$ + 1" characters, its rhs G-transformation (even leaving out of account the predicate a in the composition) would be many multiple powers greater. This counting inconsistency is detailed in reference (Ferreira 2008, 2), though misunderstood to lead to an infinity described by their Theorem 2.1

> *The Gödel's number of the formula* G *is not a finite number.*

rather than as a result of G-numerical1 numeral misplacement.

Mr. Hoge does not reply, though at a website he maintains, we find (Hoge, Supernatural 2010):

> In response to this essay, several have challenged the validity of the interpretation of *g* as a statement that begins an endless reference on the grounds that *g* and *g'* are identical statements. They have argued that since *g'* is simply the Gödel number of *g* and is identical in syntax to *g*, it is really the *same statement* as *g*, and that *g* should be interpreted as self-referential, not endlessly referential.[30]

Mr. Hoge presents an honest work to a forum of experts who dismiss it without careful consideration, or from lack of understanding; he has chosen to stick to his guns and we say good for him.[31] We only caution that the matter of G-numerical2 aside, clear distinction must be maintained between semantic reference and material implication. While $g \Rightarrow g'$, yet *g* does *not* refer to *g'*, and what precisely it does refer to may be read directly from the argument of the function $a(diag(k))$. The only statement in this business that refers to *g* is the meta-P statement "*g* is an a", and it is only by the *combination* of implications and reference (i.e., of syntactic and semantic inferences)

$$P_{fp} \Rightarrow P \Rightarrow \text{meta-P}' \Rightarrow \text{meta-P} \to P_{fp}$$

that *g*, as a fixed-point, may *metaphorically* be said by some to *indirectly* refer to itself.

---

[30]   In his notation, $g' = a(G[g])$, $g'' = a(G[g'])$, etc...., where a is the "non-provability" predicate.

[31]   Though in fairness to his detractors, no-one has claimed that *g'* is Gödel number of *g*.

What then refers to g'? This is where G-numerical2 comes into play. Repeated application of the substitution $\text{diag}(\underline{k}) \Rightarrow G[a(\text{diag}(\underline{k})]$ to Diagonal Composition (the Fixed-Point relation)

$$a(\text{diag}(\underline{k})) \Leftrightarrow a(G[a(\text{diag}(\underline{k})])$$

yields the infinite chain of mutual correspondences

$$g \Leftrightarrow g' \Leftrightarrow g'' \Leftrightarrow \ldots\ldots$$

by means of which indirect self-referencing proceeds via every g to a Gödel non-decidability of "endless reference to infinity".[32]

---

[32]   Interpretations in which the "undecidability" in GI consists *precisely* in endless referencing are not limited to readings of G-numerical2. A. Butrick writes, back in 1965 (Butrick 1965, 411-414):

The Gödel formula contains a designatory function, and the claim that the number named by that function does not stand in the relation "Dem" to any other Gödel number. To determine whether what it says is true, one must determine whether the formula associated with that number is demonstrable (either an axiom or obtained from the axioms by the rules of substitution and transformation). That formula, it turns out, has the Gödel number of the Gödel formula and hence contains a designatory function and the claim that the Gödel number named by that function does not stand in the relation "Dem" to any other Gödel number. To determine whether this latter claim is true one must proceed again in the same manner, and so on endlessly, obviating the notion that any decidable claim is made at all. This is to say that the Gödel formula is not decidable or demonstrable in any system, and that in that sense it is a pseudo-sentence. A true claim may be made about the Gödel formula, namely that it is not decidable as true or false.

where in our notation

$$\text{Dem}(x, y) \equiv \text{isProofFigure}(y) \cap \text{item}(\text{length}(y), y) = x$$

so that

$$\text{provable}(x) = \exists y.\ \text{Dem}(x, y).$$

Butrick apparently means:

$$\text{adiag}(k) \Leftrightarrow \text{adiag}(k) \text{ is an } a$$
$$\Leftrightarrow [\ \text{adiag}(k) \text{ is an } a\ ] \text{ is an } a$$
$$\Leftrightarrow [\ [\ \text{adiag}(k) \text{ is an } a\ ] \text{ is an } a\ ] \text{ is an } a$$
$$\Leftrightarrow [\ [\ [\ \text{adiag}(k) \text{ is an } a\ ] \text{ is an } a\ ] \text{ is an } a\ ] \text{ is an } a$$
$$\vdots$$
$$\vdots$$

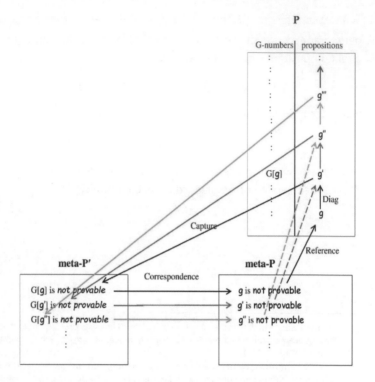

There is of course no such referencing, and LauLuna's self-referencing circle does indeed close. Though not by specification of what $a(\text{diag}(\underline{k}))$ is about, but by correct evaluation of the Fixed-Point relation that in particular avoids G-numerical1&2.

We return for a final look at the first forum page, where after an unpromising start Mr. Jones counters Smith's opening dismissal:

> The naming of the Gödel sentence does not impose arithmetical sense and behaviors upon the sentence itself, but when Gödel proceeds to treat the Gödel name arithmetically then arithmetical sense and grammar is imposed on the sentence, irrespective of what concepts the sentence employs. So Gödel employs two different calculi. The proof proceeds grammatically through one calculus, and the other calculus proceeds stepwise with it, matching it exactly.
>
> The proof therefore applies a procedural occasionalism and is so-called because it requires the intervention of a mathematician to keep the two calculi in step with each other in seeming agreement. As long as the two calculi are kept in step the grammar appears correct even though the sense is incoherent. Another example

of occasionalism is where a god is required to keep in step the disparate, non-interacting substances of mind and matter in a, seemingly, seamless interaction.

There have been many attempts since Gödel to improve on the coherency of the picture out of which his Incompleteness derives. One approach mimics the involved languages, or calculi, by natural languages whose grammars are so naturally distinct (FitzPatrick 1966). Some will find the picture more compelling than others.

Gödel, of course, features only two languages (P and meta-P), while we here display all four essential to his method. Other reviews of GI sometimes feature three languages; the choice beyond the minimum two is largely open .

Given a meta-P confined to syntactic statements about P, there is also the language of the analysis itself, the "proof language"—the language in which Gödel reasons and in which we are now speaking. In addition to statements about P and meta-P, it includes the statements themselves: it is the metalanguage of proximate higher rank in Tarski's hierarchical conception of metalanguage (Hodges 2018)–a meta-meta-language, if you like.

Here, as mentioned in our introduction, for the purpose of exposition we have partitioned off parts of the proof language, P' and meta-P', though we could have parsed along different lines had we wished a different emphasis.

GI itself, along with most other correspondences here, appears only in the proof language, where too numerical co-references

$$x = G[a]$$

are sustained and read as "x is the G-number of a". Only there does the concept "G-number" appear at all, and the diag functional relation

$$diag(G[a]) = G[a(G[a])]$$

is understood to give Diagonal Composition

$$adiag(k) \Leftrightarrow a(G[adiag(k)]).$$

The sentence g is a statement in the formal system P, interpreted as the theory of natural numbers ($\mathbf{N}, +, \cdot$). An analysis in which P is left un-interpreted and an interpreted copy is partitioned off (essentially our P) would seem better suited to address the concerns brought up by Mr. Jones. Let us call this part of the proof language Arithmetic (A), leaving in an uninterpreted P a g that says nothing at all (Lacey and Joseph 1968). In that case, GI becomes

$$g_A \Leftrightarrow g \text{ is not provable}$$

which, given the syntactic identity of g and g_A gives,

$$g_A \Leftrightarrow g_A \text{ is not provable}.$$

Whether this approach conveys the clear sense we hope, while also guarding against mathematic occasionalisms as we think it does, we are happy to leave as an open question. Meanwhile, it is probably safe (with Russell) to continue teaching that 3+5=8.

What does g not say?

A final word. Smith's opening remarks addressed to Jones above are strictly in line with what we say on the subject in the earlier section, What is$Var(x)$ says. Elsewhere Smith writes (Smith 2013, 159):

> The wff g is just another sentence of P's language, the language of basic arithmetic, LA. It is an enormously long wff involving the first-order quantifiers, the connectives, the identity symbol, and 's', '+' and '×', which all have the standard interpretation built into LA. And on that interpretation, g is strictly speaking a complex claim about the results of ordinary arithmetical operations, no more and no less.

This matter could not be put more squarely. A few lines later, however, the reader learns that in some "limited sense" g also *says* "I am not provable", and is left to wonder how this could be the same g.[33]

One might speculate how anyone could come to this. . As rightly said, as with is$Var(diag(x))$, and indeed is$Var(x)$, it is the sheer enormity of the string-length of the statement g in P that for practical matters calls for some manner of formal denotation, such as g. Unlike "3+5=8"[34]

---

[33]   In fairness, this is not a subject that Smith properly treats, and our criticism does not count much against his clear exposition of GI. He is clearly uncomfortable with this usage and rightly takes pains to attribute the declarative directly to Gödel himself.

[34]   On the material of the primitive recursiveness of addition, Gödel's Theorem V, and the Soundness of P.

$$3+5=8 \Leftrightarrow 3+5=8 \text{ is provable}$$

however, whose sense could hardly be mistaken, the blind ascription, g, carries no such overt sense, and its appearance in the material identity

$$g \Leftrightarrow g \text{ is not provable}$$

might inadvertently be mistaken as such an attribution.

Gödel himself essays in his introduction:

> The analogy between this result and Richard's antinomy leaps to the eye; there is also a close relationship with the "liar" antinomy,[35] since the undecidable proposition [R(q); q] states precisely that q belongs to K, i.e. according to (1), that [R(q); q] is not provable. We are therefore confronted with a proposition which asserts its own unprovability.[36]

which same proposition by the 1950s is given to speak demonstrably (Lucas 1961)

> "*This* formula is unprovable-in-the-system"

by the postmodern 1960s speaks in first person (Thomas 1995)

> "I am not provable"

and by the dawn of the new Millennium, proclaims of itself (Hofstadter 1999)

> "I am not a Martian Number"

and other wondrous things.

If nothing else, one has to wonder at so many apparent paraphrasings of this single denotation g:

> ...this formula is unprovable; this statement is unprovable; G cannot be proved within the theory T; I am improvable; I am unprovable; I have the property expressed by the formula g; I am unprovable in PA; I am not provable from A; This wff cannot be proved in S; g cannot be proved; g is unprovable in K; this statement is unprovable; g is unprovable in the theory; G is PA-unprovable; My code has prop-

---

35   http://www.geier.hu/GOEDEL/Godel_orig/godel3.htm#14
36   http://www.geier.hu/GOEDEL/Godel_orig/godel3.htm#15

erty g; there is no number that is a proof figure for g; there is no proof of this formula; I am not demonstrable; This formula is unprovable-in-the-system; There is
no proof for proposition G in the system X; etc.

and why they are so rarely introduced by their authors without caveat:

...says in some limited sense (Smith 2013), may be regarded as expressing (Gödel
1962), essentially says (Contributors, Gödel's Incompleteness theorems 2021),
states in other words (Gödel 2000), asserts in a roundabout manner (Mangin 2002),
says in the eye of P (Lajevardi and Salehi 2021), intuitively asserts (Boolos, Burgess
and Jeffrey 2002), basically says (Askanas 2006), says in one interpretation (Suber,
GI 2002), says essentially (Suber, GI 2002), interpreted within Gödel's system P,
states (Lipscomb 2010), is equivalent in the theory to the statement that asserts
(Boolos 1994), translates into (Lucas, Conceptual Roots of Mathematics (Chapter
8) 2002), intuitively says (Marker 2006), may be taken as making the claim (Butrick
1965), could be taken as saying of itself (Butrick 1965), says in effect (Lucas 1961),
has the same form of the proposition (Thomas 1995), is indistinguishable from the
proposition (Thomas 1995), etc....

On the other hand, what great harm can they do, these readings of g? To
the first Incompleteness theorem? None. To the second? None. They are
innocuous, benign... if nothing else, a bit of harmless fun. What difference
then does it make how we read g?

The difference comes at the speculative margins of GI, e.g., when
the Turing Machine is construed as a model of P in a mind vs. machine
argument. Consider philosopher Hillary Putnam's reading of the second
Incompleteness theorem (Putnam 1979, 366):

Given an arbitrary machine T, all I can do is find a proposition g such that I can
prove:
(*) If T is consistent, then g is true and undecidable in T.
However, T can perfectly well prove (*) too! And the statement g, which T cannot
prove (assuming consistency) I cannot prove either (unless I can prove that T is
consistent, which is unlikely if T is very complicated).

whereas the actual T-syntax of the (*) conditional reads:

$$\neg \; prov(n) \rightarrow \neg \; prov(m)$$

when $n=G[0=1]$, and $m = G[g]$ (Smith 2013) (Marker 2006).

One need only observe that Putnam's system T cannot prove what
may not be stated in the language of T (any more than a Lucas-Penrose
system (Lucas 1961) may prove its contrary), and that what can be stated
in T is constrained by the possible models, formal interpretations, of T—

such statements as demonstrate the deductive power, or *reasoning*, of a machine from a set of initial premises (e.g., axioms over some domain of discourse given upon formalization), and that are not merely natural language sentences assigned to fully-formed arithmetic strings, "perroquet parler".[37] The constraint does no harm to either first or second Incompleteness theorems, and gives one to understand them, it would seem, more correctly. If nothing else it may help preclude a good deal of idle GI speculation.

There is yet another sense, independent of semantic containment, in which it is sometimes claimed that *g* asserts its own non-provability. The argument proceeds via a supposed isomorphism of the meta-language and appears to be favored by non-mathematicians. As David Wayne Thomas explains in his award-winning *Gödel's Theorem and Postmodern Theory* (Thomas 1995, 248-261), "*Gödel's theorem* shows that there can always be *propositions that signify* in language and metalanguage simultaneously and that they can be constructed to signify at *cross-purposes*."

Accordingly, as the false-friend (Contributors, False friend 2021)

art

signifies differently in English than it does in German, the argument goes, so the expression denoted *g* signifies differently in P than it does in meta-P. By this reasoning then, while the expression denoted is$Var(x)$ signifies in P arithmetically, in meta-P it is read "$G^{-1}[x]$ is a variable".

Upon outlining the G-transformation, Gödel himself speaks directly to the question of the metalinguistic manner of signification (Gödel 1962, 46):

> We represent by the same words in italics those classes and relations of natural numbers which have been assigned in this fashion to such previously defined metamathematical concepts as "variable", "formula", "propositional formula", "axiom", "provable formula", etc. The proposition that there are undecidable problems in the system P would therefore read, for example, as follows: There exist *propositional formulae a*, such that neither *a* nor the *negation* of *a* are provable formulae. (Gödel 1962, 46)

---

[37]   Fr. parrot-speak.

Thus, the metalinguistic syntax is, as usual, that of ordinary language, and it is the usual sign of informal syntax, where "x is a *variable*"—one of the above referenced R' classes of natural numbers—that signifies "x is a variable in P". Thus, while the expression isVar(x) also appears in meta-P (in the usual, assumed, Tarskian metalanguage (Gómez-Torrente 2019)), it is read there just as it is read in P, arithmetically.

This false-friend conception of *g* is also to be found in David Hofstadter's "Gödel" article for *TIME* magazine's widely circulated *100-Most Influential People of the Millennium* edition (Hofstadter 1999), as well as in his Pulitzer Prize winning book, *Gödel, Escher, Bach* (GEB) (Hofstadter 2000). But in addition to its false-friendliness of *g* there is also a kind of "translation", a kind of "isomorphism", a kind of "transcription", a kind of "embedding", a kind of "double entendre" (etc), of some statement in P, and one has to wonder at the author's intent. When we consider a typical context in which these and like predications occur:

> Stepping out of one purely typographical system into another isomorphic typographical system is not a very exciting thing to do; whereas stepping clear out of the typographical domain into an isomorphic part of number theory has some kind of unexplored potential. It is as if somebody had known musical scores all his life, but purely visually—and then, all of a sudden, someone introduced him to the mapping between sounds a musical scores. What a rich, new world! Then again, it is as if somebody had been familiar with string figures all his life, but purely as string figure devoid of meaning-and then, all of a sudden, someone introduced him the mapping between stories and strings. What a revelation!

and given the subtitle of the work[38] it would seem safe to take this usage figuratively, idioms in the author's poetic flight.[39] The *100-Most* article

---

[38]   *A Metaphorical Fugue on Minds and Machines*

[39]   Or if one wished to take it more seriously...
Briefly, to the extent that it is a 'meaningful' (page 59) "double entendre" (page 60), in the author's own non-standard diction, *g* derives as a member of a "higher level isomorphism" (page 58) which "completes" (page 271, 272) some "lower level isomorphism" (page 58) of a formal deductive system (i.e., of its alphabet, signature), that has two "passive meanings" (page 59). Simply, it is a formula in a formal deductive system that is modeled by two structures. (To the same extent then, all formulas in P, e.g., "3+5=8" are double entendres; cf. footnote(24).), or more formally (Beziau 2007)
Two logical systems L_1 = and $\langle \Sigma_1, \vdash_1 \rangle$ and L_2 = and $\langle \Sigma_2, \vdash_2 \rangle$ are isomorphic if and only if there exists a signature isomorphism h : $\Sigma_1 \to \Sigma_2$ such that $h[\Phi^{\vdash_1}] = h[\Phi]^{\vdash_2}$ for every $\Phi \subseteq L_{\Sigma_1}$.
g eventually appears as part of the higher isomorphism on formal system 'TNT' given

too, colorful as it is, lacks sufficient structure to account for Gödel's In-completeness mechanism: there are too few "players" to pull it off.

Briefly, the Martian alphabet corresponds (by pure coincidence) to what are numeral digits on Earth, 0-9, so that the theorems they translate from Earth-mathematics text books only look like huge numerals to the puzzled Earth mathematician. It seems however that they are not entirely random, and the Earth mathematician eventually develops an elaborate mathematical theory about what numerals may or may not be found in Martian textbooks: "Martian Numbers". When one day an ingenuous Earth-mathematics student derives a theorem

X is not a Martian Number

in which "X" is precisely its Martian translation, which, despite its correct-ness, could never appear in Martian mathematics books, assuming their adequacy.

Now, while Earth mathematicians might propose equalities of sums and multiples of natural numbers, and so forth, they are in no position to propose what happens on Mars. That might be left to a Jupiter mathema-tician, who'd speak of them, when deigned, in a superior tongue—a higher language—as a Pluto mathematician might likewise of the goings-on on Earth.

Enter into this happy arrangement an advanced extra-solar being, an Alpha Centauri Kurt Gödel or a David Hofstadter. Only *He* sees all. He

---

on page 276 (and which is only partial); this is the TNT version of Gödel's 45 PRR/meta-P' correspondences, our Capture. With it, however, there appears no cor-responding lower order isomorphism of the signature of P, *and there is none availa-ble.*

"But what is an 'option'?" the man asks, to which the automated attendant at the other end of the phone line responds, "I'm sorry, I did not understand you; will you please repeat your answer?" but it hopeless; indeed, she does *not* understand. So it is with higher level statements that lack lower level foundation. While the deductively reasoned output that constitutes what the attendant *says* is quite a lot and good (in the language of Boolean Arithmetic), what we hear, correlated to some selected por-tion of it, is "parrot speak".

The same may be said with respect to the meta-P *reading* of the diagonal lemma

$$adiag(k) \Leftrightarrow a(G[adiag(k)])$$

as it appears in P

$$adiag(n) \leftrightarrow a(m)$$

(where *in meta-P*, $n = G[adiag(x)]$, $m = G[adiag(n)]$) in course of the so-called P-for-malization of the first Incompleteness theorem. The reading is not accomplished via "definitional extension" (Smith 2013) of the P-signature by letter 'G'.

sees that when Martians say "*g*", Earthlings hear no proposition at all, but only an interminable trailing off of numeral digits: "Thirty-centillion, two-hundred twenty-seven nonillion,...", i.e., "G[*g*]". He sees that hearing this, Jupiterians and Plutonarians persist in their respective linguistic studies, one isomorphically re-inventing the semiotic wheel of the other. Then correlating these findings into a sequence of correspondences, he (as only he, seeing all, can) derives the curious self-implication

$$g \Leftrightarrow g \text{ is not provable}$$

that to him seems to tell of a kind of Incompleteness—of a futility inherent to the Earth-born project that presumes to formalize axiomatic systems of Arithmetic. *He* sees all, and laughs.

III. An Odious Turn

In the last section we have seen the dangers of misreading GI, and the ease with which one falls victim to those dangers when ignoring details of the derivation. It should be taken as a cautionary preview of the interpretive dangers that still lie ahead for the student of Gödelian Incompleteness who has managed to follow our derivations up to now.

In this section we introduce the additional layer of deductive complexity that appears in Gödel's actual derivation. This is the "odious turn" of its title, which in turn introduces additional room for interpretive error. In the literature this often concerns GI's relevance to questions or foundational claims concerning the reality of numbers and/or the truth of statements and equations about them.

Unlike the supernatural Gödelian numbers of Mr. Hoge or Mr. Jones' monadic Gödelian self-reference ideas of the previous section, the *foundational* claims that we find in the coming subsections are made on the authority of expert testimony, and so may present an additional layer of logico-mathematical difficulty for the GI student.

Were we to now pick up where we left off at the end of the previous "Incompleteness" section with a diagonalized negation of Gödel's 46th Capture in hand

$$g \Leftrightarrow \neg \vdash g \qquad \text{(Diag)}$$

and the reasonable assumption that the five Peano axioms give an adequate account of the natural numbers, deduction also being valid; i.e., the assumption that our deductive system P is sound

$$\vdash g \Rightarrow g \qquad \text{(S)}$$

- with these, we would swiftly proceed to bring the Incompleteness story to a conclusion

(a)

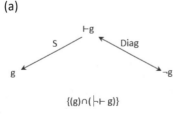

$$\{(g) \cap (\vdash \vdash g)\}$$

that "g" is a true proposition in the language of P that is not deducible from the logic of P, the Incompleteness of P, and with a final application of the soundness assumption (in contraposition)

$$\{(\neg \vdash g) \cap (\neg \vdash \neg g)\}$$

that "g" is a proposition in the language of P, that can neither be proven nor disproven by the logic of P, the Undecidability of P.

Gödel too could have proceeded in this way, but instead embarks on an elaborate sequence of deductions that after several pages omitting any mention of P-soundness brings us again to the same two results, but their deductive order reversed.[40] The heuristic offered by Gödel in his introduction, like our counterfactual (a), assumes P-soundness[41]

(b₁)

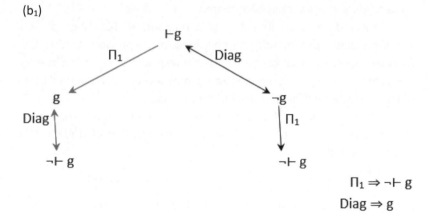

$$\Pi_1 \Rightarrow \neg \vdash g$$
$$\text{Diag} \Rightarrow g$$

---

[40]   Though deducing not exactly undecidability from Incompleteness, nor even Incompleteness *from undecidability*, but deducing rather a related though lesser-known *insufficiency* of Hilbertian-adequacy, which we discuss in subsection "Konigsberg, 1930", below (e.g., paragraphs, "The Hilbert school..." and "Gödel offers two..."). It is also there where we lay out Gödel's own derivational mechanism for the diagonalization of negation of the provability predicate (Recall opening paragraph in subsection *Fixed-Point* above.)

[41]   Said by some to be the only "mistake" in the paper (Helmer 1937, 58-60).

(b₂)

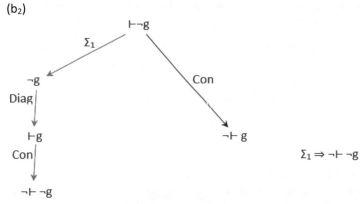

(Con ≡ consistency, Π_1 ≡ Π_1-soundness, Σ_1 ≡ Σ_1-soundness)

and serves as the blueprint for his eventual derivation, which does not. What it does have however is a deductive structure that allows for elimination of the soundness assumptions[42] by alternative syntactic deductions (in the first instance, b_1) and formal redefinition (as in the second, b_2)—a structure that (a) lacks.

Königsberg, 1930

Why would Gödel go to such trouble to eliminate the soundness assumption from his derivation? There are at least two plausible reasons, the first (and better known) as Peter Smith explains (Smith 2013):

> Hilbertians, then, thought that the status of most mathematics is to be sharply distinguished from that of some small central core of real, true, elementary arithmetic. Most mathematics is merely ideal: we can talk about what demonstrably follows from what within the game, but shouldn't talk of the statements made in the game as being true. Partly for that reason (as he later put it in a letter), when Gödel published his Incompleteness paper
>
> ... a concept of objective mathematical truth as opposed to demonstrability was viewed with greatest suspicion and widely rejected as meaningless. (Gödel, 2003a, p. 10, fn. c)
>
> Which probably explains why—as we've remarked before—Gödel in 1931 very cautiously downplayed the version of his Incompleteness theorems that depended

---

[42]   Subscribing to a *Computability* arithmetical-hierarchy (Soare 2016) (Moschovakis 2014) (Sutner 2017) (Haskel 2015) in which a primitive recursive R(x) is Δ_0, $\exists x.R(x) \in \Sigma_1$, and $\forall x.R(x) \in \Pi_1$—a hierarchy we think most suitable to Gödel's own analysis of Incompleteness.

on the assumption that the theories being proved incomplete are *sound...* (Smith 2013, 274)

The Hilbert school of mathematics of the mid-1920s had been instituted and purposed on an investigation into the foundations of mathematics— a study not of numbers themselves, but of number deduction systems, number theories, and their suitability or "adequacy" to the job, this being their soundness. While the soundness of the finite classical arithmetic relations, "Surveyable in all its parts", that constitute so-called Real Arithmetic, was agreed by all to be soundly "unproblematic", Brouwer and the Intuitionists cautioned Hilbert Formalists to "refrain from attributing non-intuitionistic classical mathematics a material meaning"—i.e., from attributing content to the non-finitistic arithmetic relations that constitute so-called Ideal Arithmetic.

However meaningful or meaningless Ideal arithmetic relations, Hilbert could not bring himself to wholesale discount, as instruments, their use in deductions practically unattainable by the limited finitistic methods of Real Arithmetic,[43] though he recognized that some account ought to be made of their suitability. It would certainly be significant e.g., were Ideal mathematics a consistent extension of Real mathematics, though signifying precisely what? Perhaps adequacy of the whole. This is the gamble Hilbert takes, the pursuit of which i.e., the proof of such a consistency, assumed adequate, being a centerpiece of his program.

Naturally, not all in the business were on board, for example Brouwer's Intuitionists. It was they, after all, who first called into question this extension of number intuition to Ideal infinities. But they were not alone. For Frege, unlike his former student Wittgenstein, the mathematician is not a creator, but a discoverer of mathematics, whose axioms he presumes to formalize must first *refer* to actual states of affairs, mathematical facts; they must first be true. Gödel too understood mathematics as a descriptive science: its subject, numbers, existing independent of what might be said of them (Feferman 1997). For even as formalization may improve upon the accuracy of his account, Gödel thought, the mathematician's job remains discovery and description of extant number rela-

---

[43]   Hilbert's foundational "views" were in fairness (and appropriately) a forever evolving work in progress (Zach, Hilbert's Program/1.1 Early work on foundations 2019).

tions, not the creation of them. In the weeks that followed his momentous discovery, which seemed to support this philosophic pre-disposition, the closet Neoplatonist would nonetheless have to tread a careful course, mindful of a thousand things, possibly including his fledgling professional prospects.[44]

But as patience is not a virtue of youth, a discussion session on the Foundations of Mathematics in 1930 attended by Gödel presented a close call. The anchoring symposium was being held in nearby Königsberg (Dawson 1984) (Feferman 1986) just weeks before Gödel was to publish his Incompleteness paper, though even then he had the proof in hand. And so, increasingly bored—annoyed at having to listen to so much of which he knew better[45]—and egged on by von Neumann (also annoyed), Gödel decided to give the distinguished Königsberg audience a preview of what anyway was soon to be known by all. Earlier that day Heyting had spoken on behalf of the Intuitionists, Carnap for the Logicists, while before them now stood a relatively unknown Privatdozent saying exactly who knows what.

Having announced his result, while offering little more than an outline of his method, Gödel took his seat to uncertain, polite applause. Perhaps the young man had said something of import, perhaps not, but lunch was being served at 1... Fortunately Hans Hahn, Gödel's thesis advisor, was also there and could quickly moderate the premature announcement,[46] though had any in attendance grasped its significance as well as Von Neumann had, Gödel would likely have faced serious criticism from many sides.

While there was not as much opposition to Gödel's eventual publication as there might have been, that such caution had been warranted would soon be apparent. Among logicians of standing, only Zermelo openly disputed Gödel's findings (Dawson 2005). The two men first met at the annual meeting of the German Mathematical Union held in Bad Elster, 1931, and on friendly terms. In the brief correspondence that fol-

---

[44]   E.g., Smith, above (Smith 2013) (section 37.2) cites Feferman's 1984, *Kurt Gödel: conviction and caution* (Feferman 1997).

[45]   In particular, of Hilbertian Adequacy.

[46]   Perhaps on the instruction of his academic advisor Hans Hahn—an old hat, Feferman suggests (Feferman 1997) page 102), who knew how to bide his time.

lowed however Zermelo would eventually express doubts specifically regarding the veracity of Gödel's method of definitional Captures. In at least one such exchange (Dawson 1985) he speculates that were one to Capture contentual correctness rather than Provability, there would derive "a contradiction analogous to Russell's", clearly not having understood the derivation. And though Gödel took great care to explain in his reply, answering such objections (with a 10-page reply in this case), the following year Zermelo would publish a "scathing dismissal" of GI (Zermelo 1932).

The more dangerous threat was posed by sometime Logicist, and sometime Formalist, Ludwig Wittgenstein, renowned author of the *Tractatus Logico-Philosophicus* (Wittgenstein 1922). For the received assessment of his views on Gödel's Incompleteness, we quote a passage from Dawson's *The Reception of Gödel's Incompleteness Theorems* (Dawson 1997) reviewing Wittgenstein's *Remarks on the Foundations of Mathematics* (Wittgenstein 1978):

> Several commentators have discussed Wittgenstein's remarks in detail (see, for example, the articles by A. R. Anderson (1958), Michael Dummett (1959), and Paul Bernays (1959), and nearly all have considered them an embarrassment to the work of a great philosopher. Certainly it is hard to take seriously such objections as "Why should not propositions.. .of physics...be written in Russell's symbolism?"; or "The contradiction that arises when someone says 'I am lying'...is of interest only because it has tormented people"; or "The proposition 'P is unprovable' has a different sense afterwards [than] before it was proved" (Wittgenstein 1956, pp. 432 and 434). Whether some more profound philosophical insights underlie such seemingly flippant remarks must be left for Wittgenstein scholars to debate; suffice it to say that in Gödel's opinion, Wittgenstein "advance[d] a completely trivial and uninteresting misinterpretation" of his results. (Gödel to Abraham Robinson, July 2, 1973).

the gist of the more direct complaint against GI that appear in a *Notorious Paragraph* (Floyd and Putnam 2000) (Lampert 2006) later in the Remarks being that there could be no *truth* to a P-proposition outside its *provability* in P, making of the Incompleteness claim then a nonsensical self-contradiction. We conduct a more thorough discussion in our final section "Wittgenstein's Perspective" below. As most commentators agree, fortunately for both men Wittgenstein's *Remarks* were only published posthumously (1956), long after Gödel's work had gained a more general understanding and acceptance.[47]

Gödel himself, of course, lays no claim to the formal proof of any truth, and word eventually spread that in preparation for the review Wittgenstein had read no further than the aforementioned introductory heuristic, a rumor that persists to this day (Feferman 1998). Faced with the prospect of entering the losing side of a Frege-Hilbert controversy 30 years in running, or the same with respect to the one between Brouwer and Hilbert, it is perhaps little surprise that even as the uniformity of Real Arithmetic seems to straightforwardly grant it (see e.g. footnote (54) following diagram (B) below.), Gödel elects to forgo the convenience of a semantical Π_1-soundness assumption in favor of lengthy, grueling "pure syntactics".

Another possible reason that Gödel might have gone to such trouble to steer clear the soundness assumption quite apart from all this is that Incompleteness may not have been his main aim, nor even Undecidability. At the completion of the Undecidability proof in the original paper, there immediately follows an unnamed informal and seldom cited corollary, which has the Hilbertian empirical-adequacy condition (i.e., Consistency) insufficient to its purpose. While in presentation the corollary follows Undecidability, deductively, it follows Incompleteness: Given

$$\text{(A)} \qquad \text{Con(P)} \Rightarrow \neg\vdash g$$

and

$$\text{(B)} \qquad \neg\vdash g \Leftrightarrow g$$

were we now to amend $\vdash\neg g$ to the axioms of P, its $\tilde{P}$ extension would remain consistent

$$\tilde{P}: \qquad (\neg\vdash g)\cap(\vdash\neg g)$$

---

[47] The more recent publication of Wittgenstein's *Nachlass* (Wittgenstein 2001) places the *Remarks* in a wider context and perhaps suggests a more nuanced reading (Shanker, Wittgenstein and the Turning Point in the Philosophy of Mathematics 1987) (J. Floyd 2001). It is in any case interesting how Zermelo and Wittgenstein both make the same proof (heuristically and inadvertently) that Gödel and Tarski only later make of the Undefinability of Truth *in a consistent P*.

though unsound

$$\tilde{P}: \qquad (\vdash\neg g)\cap(g)$$

and so "Inadequate", the consistency condition proving *insufficient* to its purpose.

It is a proof that assuming adequacy itself (i.e., soundness rather than consistency) of course rules out.[48] And let us recall that this is the very insufficiency that Gödel (with Von Neumann's support) points out at Königsberg:

> ...it is quite possible that one could prove a statement of the form (Ex)F(x) where F is a finitary property of natural numbers (the negation of Goldbach's conjecture has this form, for example) by the transfinite methods of classical mathematics, and on the other hand could realize via contentual considerations that all numbers have the property not-F; indeed, I would like to point out that this would still remain possible even if one had demonstrated the consistency of the formal system of classical mathematics.

so that

> if one adjoins the negation of such a proposition to the axioms of classical mathematics, one obtains a consistent system in which a contentually false proposition is provable.

I.e.,

> A statement provable in a consistent formal system might nonetheless be false.

two months prior to publishing the Incompleteness paper.[49]

Given then, $Con(P) \Rightarrow \neg\vdash g$, from (A) below, what conditions in place of Hilbertian adequacy *might be sufficient* to prevent extension of this P to an inadequate $\tilde{P}$ via axiomatic amendment $\vdash\neg g$?

Gödel offers two alternatives: The semantic Σ_1-soundness

$$\neg[\, g \cap (\, \vdash\neg g\,)\,]$$

and another "purely syntactic" condition of his own creation. For this we first define the primitive recursive "R(x)"

---

[48]   The same may be said of Gödel's also informal GI-2 "proof" a couple paragraphs later... i.e., that proof too forestalled on such an assumption.

[49]   The full transcripts of the "Discussion" are only now gaining wider circulation (Dawson 1984) (Dawson 2005).

$$g \Leftrightarrow \neg prov(G[g]) \quad [= \forall x.\neg proves(x, G[g])$$
$$\equiv \forall x.R(x) \qquad]$$

and the square bracket notation we shall use as shorthand for its meta-linguistic universal generalizations:

$$[R] \equiv R(n), \text{ for any } n$$
$$[\vdash R] \equiv \vdash R(n), \text{ for any } n .$$

Gödel's second purely syntactic condition, which he calls *ω-consistency*, is then defined

$$\neg[[\vdash R] \cap (\vdash \neg g)].$$

These two conditions, one semantic, one "purely syntactic", are both offered at Königsberg as alternatives to insufficient Hilbertian adequacy. Undecidability then[50] might well have been only of Gödel's secondary interest, considering the emphasis he placed on them at the time.

In a now famous 1904 address to the 2nd International Congress of Mathematicians, Hilbert posed a number of important unanswered mathematical questions that came to be known as "Hilbert Problems" upon his conviction that "in mathematics there is no *ignorabimus*" (Feferman 1998, 3-28).[51] One of which, the "Decision Problem", inaugurated on its own a family of such decision-problems, including what would eventually become Gödelian decidability. By Gödel's time however these were of little interest to the up-and-coming mathematical logician, including Gödel, mainly because none were in the foundational focus of the-then dominant Hilbert Program[52] of the mid-1920s, whose putative purpose had been to determine once and for all the parts of mathematics that were deductively secure, and in particular, in compliance with Intuitionist complaints against Hilbert's earlier non-finitary allowances.

---

[50]   The "First Incompleteness Theorem" proper, while syntactical Incompleteness itself, again, Gödel does not even mention.

[51]   As a counter to Reymond's declaration *"ignoramus et ignorabimus"* (Contributors, Ignoramus et ignorabims 2021) 20 years earlier.

[52]   As Zach's Stanford Encyclopedia article (Zach, Hilbert's Program/1.2 The influence of Principia Mathematica 2019) relates, "Nevertheless, other fundamental problems of axiomatics remained unsolved, including the problem of the 'decidability of every mathematical question,' which also traces back to Hilbert's 1900 address." See also lengthy Hilbert quotation in Feferman, *Deciding the Undecidable.pdf* (Feferman 1998), page 15 of the Appendix.

Those taking part in the Program then hoped to prove the empirical adequacy of the whole of Classical Mathematics (including non-finitary, Ideal parts, accepted non-contentually) assuming its soundness to follow from nothing more than its self-consistency (Zach, Hilbert's Program/1.2 The influence of Principia Mathematica 2019) (Rathjen 2009). Only a year after the question had been posed by Hilbert and Ackerman in the context of the Program, Gödel proved the Completeness of first-order logic.[53] Meanwhile in pre-Nazi Warsaw, one of Alfred Tarski's students, Mojzesz Presburger, had the year before made significant progress on an older decision problem of Hilbert Problems fame, proving the same for a significant fragment of Arithmetic (Presburger 1929). While both results could be said to have advanced the prospects for the Program, they differed in scope, Gödel's being more central to it.[54] In any case, this mounting good news could well have called into question or at least obscured the clear message Gödel had hoped to send regarding the soundness of the Program. What better way to avoid this prospect, while adding to the plausibility of Hilbertian Insufficiency, than to clarify by demonstration the distinction between the two decision problems? For if Undecidability, as a demonstration that Presburger's earlier near-miss is as near to Arithmetic completeness as can be realized—if that too might be obtained along the way, and at little expense—then why not?! This is the narrative of Gödel's thinking we think best fits the available historical and deductive facts. Undecidability, with its ω-consistency pretexts, accordingly, we take as only incidental to Gödel's main line.

Gödel begins down this *syntactic way* assuming the consistency of P, $\vdash\varphi \Rightarrow \neg\vdash \neg\varphi$, and his Theorem V, which with only a sketch of a proof claims the Δ_0-completeness of P:

$$[R(n) \Rightarrow \vdash R(n)], \text{ for any } n \quad V$$

so that in our notations

[53] $\models\varphi \Rightarrow \vdash\varphi$, for P without the Peano axioms.

[54]   Presburger's promising result (Haase 2018) (Feferman, The impact of the Incompleteness theorems on mathematics 2006) (Zygmunt 1991), answering in the positive for at least a significant fragment of arithmetic (P without the multiplication operation), though predictably meeting with lackluster response. While encouraged by Presburger's result, the deductive completeness of Arithmetic had by then ceased to be one of Hilbert's main interests and was never incorporated into the new Program.

⊢g   ⇒ [⊢R]   UI (algorithmic UI)
[R(n)]   ⇒ [⊢R]   V
G   ⇔ [R]   UG
¬∃x.R(x) ⇔ ∀y.¬R(y)   (i) algorithmically, uniformly [55]

by means of which finer-markers we parse the above b_1 and b_2 semantic-UI trees:

(A)

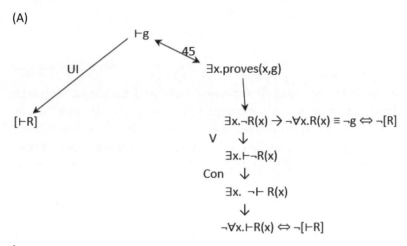

I.e.,

$(⊢g) → ⊥$

By Modus Ponendo Tollens [¬(p∩q) ⇒ (p→¬q)] , assuming P-consistency, then

∴ ¬⊢ g

I.e., the last by universal 'y' uniformity over R(y) → ∃x.R(x)

(B)

$$\neg\vdash g \Leftrightarrow \neg\exists x.\text{proves}(x,g) \Leftarrow\Rightarrow \forall x.\neg\text{proves}(x,g) \Leftrightarrow [R] \Leftrightarrow g$$

$$\Downarrow \qquad\qquad V$$
$$[\vdash R]$$
$$\Downarrow \qquad\qquad \Delta_0$$
$$[R] \Leftrightarrow g$$
$$\Downarrow \qquad \Sigma_1 \qquad\bigg\} \omega$$
$$\neg\vdash \neg g$$

56

$$(\Delta_0 \equiv \Delta_0\text{-soundness})$$

Notice that in obtaining (B) above from our earlier $b_{1\&2}$, we substitute, at places, *alternate syntactic deductions* for the unwanted soundness assumption (when the soundness is Π_1), while at others where this cannot be done (where the soundness is Σ_1) we absorb it into the "purely syntactic" rule-of-inference, ω-consistency—in fact, a purely cosmetic formal definition

$$\begin{array}{ccccccc} \Delta_0 & & \text{UG} & & & \Sigma_1 & \\ [\vdash R] & \Rightarrow & [R] & \Leftrightarrow & g & \Rightarrow \neg\neg g \Rightarrow \neg\vdash\neg g & \qquad \omega \end{array}$$

whose epistemic authority rests on none other than the $\Sigma_1 \cap \Delta_0$-soundness of P.[57] Gödel thus derives Incompleteness, Undecidability, Insufficiency, and GI-2 without *explicit* application of a P-soundness assumption i.e., *by "purely syntactic"* inference.

The usage has become ambiguous, as a rule of inference that is expressed at all is expressed in a syntax, to wit, syntactically. The real question—the one this semantic vs syntactic usage more often pretends to—

---

[56]   Note however, that while $(\Delta_0 + \text{uniformity}) \Rightarrow \Pi_1$,
$$\neg[(\Delta_0 + \text{uniformity}) \Rightarrow \Sigma_1]$$
since
$$\neg[\exists x.(A(x) \to B(x)) \Rightarrow \exists x. A(x) \to \exists x. B(x)]$$

[57]   I.e., a rule "purely syntactic" by notation only. No different, it is true, for the underlying epistemic authority behind the consistency assumption itself and corresponding syntactic inference rule; there, the task of proving the consistency of a deductive system is a "relative consistency" that falls to the task of finding for it a model, as Hilbert three decades earlier does for Geometry (Blanchette 1996) (Blanchette 2017) (Blanchette 2018). Why else do we so readily accept the reasonableness of ω-consistency? It is because to prove some A(x), for any x, is to also prove that *in any model* "A(x), for any x", whether quantifiers belong to the formal alphabet or not.

is epistemic, that of the rule's *validity*, its justification. The soundness condition is in this sense given no less syntactically than modus ponens, only that while soundness does not always hold, and its truth value ascertained only via the *meaning* of its terms (and is so deductively semantic), modus ponens preserves truth, *regardless of the meaning* of its terms (and is so deductively non-semantic), and it is there—not in, "⊢" vs "T" notations—where lies the deductively significant distinction. Inclusion of "⊢" symbolism does not a non-semantic rule make.

While a soundness assumption (in particular, Σ_1-soundness) does indeed show up above, at least *epistemically*, in the syntactical ω-consistency (where it cannot be substituted by *alternate syntactic deductions*, and is instead absorbed into the *formal definition* of ω-consistency), without it (i.e., where the soundness assumptions in $b_{1\&2}$ are indeed replaced by *alternate syntactic deductions*) Gödel derives (1) Incompleteness, (2) Hilbertian Inadequacy, and (3) Unprovability of Consistency. Undecidability is itself independent of, hardly even incidental to the purposes of, Gödel's main foundational concerns, and is certainly incidental to proper consideration in the truth-deflationism discussion of our coming subsections.

Be that as it may, this clever repackaging of an unseemly semantic assumption was the final aesthetic stroke atop a work of bonafide logical genius, yet a work still accessible to the understanding of undergraduate mathematics. By the time the continental establishment had realized the implications to their beloved Hilbert Program, the young prodigy would be safely away at a prestigious American research institute, IAS.

Princeton, 1962

Feferman's axiomatics

There were nonetheless occasional, if niggling, doubts regarding the formal veracity of the proof. Given the wide acceptance of the views it challenged, and again Gödel's own youth and inexperience, it is remarkable how few doubts one finds in the literature (Feferman 2006) (Dawson 1997). What one does find is a concern, if not doubt, over whether the derivation had been "constructive",[58] as Gödel repeatedly claims it to

---

[58]   I.e., "Intuitionistically unobjectionable", particularly with regard to Σ_1 assertions.

be.[59] Another and related matter concerned Gödel's unsubstantiated assurance that the full proof could be formalized to the satisfaction of strict Formalists, who might feel his original informality to be foundationally weak (Feferman 1962).[60]

Aware of and sympathetic to such concerns, staunch GI advocate Solomon Feferman ranks Gödel's Incompleteness among the theorems of "True Arithmetic", whose deductive gaps are "bridged.... by the addition of certain non-constructive [infinitary] rules of inference" (ibid, 259), such as Hilbert's ω-rule[61]

---

[59]   On its face e.g., ω-consistency follows only from a non-constructably (lower-case) infinite set of premises—on this count too then, non-Hilbertian, though only of the later Hilbert.

[60]   Which purpose is an absolute logical rigor that "deals only with statements which we derive from explicitly stated assumptions by explicitly stated and readily verifiable rules." as Curry in *The Purposes of Logical Formalization* (Curry 1968) explains it.

[61]   We cite Wittgenstein earlier as saying that there is no *truth or untruth of a mathematical proposition* without reference to the formal system in which it is deduced, and something of the sort does indeed seem to hold as regards Gödel's two systems— g being unprovable in P, while provable (thus *meaningfully* true) in $\bar{P}$. In any case, the proviso certainly applies to the *provability or unprovability of a mathematical proposition*, and it does not take a Gödel to tell us that. Hilbert's ω-rule (Ignjatović 1994)
RULE ω*. Whenever A(x) is a quantifier free formula for which the following can be finitistically shown: A(z) is a correct numerical formula for each particular numerical instance z, then its universal generalization can be taken as a new premise in all further proofs.
for which $\bar{P}$ is the extension of Gödel's P by the rule.
What then are the options?
$[R] \Rightarrow g$
$[\vdash_P R] \Rightarrow \vdash_P g$
$[\vdash_{\bar{P}} R] \Rightarrow \vdash_{\bar{P}} g$
The first is no more than Universal Generalization, while the second would indeed accomplish Hilbert's objective in making P complete, though would also make it inconsistent; same for the third in respect of $\bar{P}$.
So what can be done? What use might be made of a rule that on one reading makes inconsistent the deductive system about which it is proposed, and on another is redundant? How might we salvage the last idea of a great and beloved man on the verge of retirement? One thing might be to revise the rule to clear this ambiguity in such a way that lends it some practical functionality. We might try:
Whenever A (x) is a quantifier free formula for which the following can be finitistically shown in *one deductive system*: A(z) is a correct numerical formula for each particular numerical instance z, then its universal generalization can be taken as a new premise in all   further proofs *in the deductive system resulting from the amendment of this rule.*
I.e.,
$[\vdash_P R] \Rightarrow \vdash_{\bar{P}} g$ ω↓

RULE ω*. Whenever A(x) is a quantifier free formula for which the following can be finitistically shown: A(z) is a correct numerical formula for each particular numerical instance z, then its universal generalization can be taken as a new premise in all further proofs.

which as stated is rather ambiguous (Ignjatović 1994) (Schmerl 1982) (Prochazka 2010).

While Gödel's ω-consistency *can* and does say

$$[\vdash_P R] \Rightarrow \neg \vdash_P \neg g$$

Hilbert's ω-rule *may not* say

$$[\vdash_P R] \Rightarrow \vdash_P g$$

without also making P inconsistent (Buldt 2014). Instead, we have in addition to and independent of the formal system an ω-rule understood, conventionally, to state something independent of what the formal system has proven.

One might well see or interpret the ω-rule as a condition that Hilbert thought a proper formal system ought to be able to satisfy; i.e., that as [R] and g refer to the same content, the truth or correctness of one entailing the truth of the other, then in a proper Hilbertian formal system the proof of one ought to entail or follow from a proof of the other—that it should in effect prove its own consistency, via soundness, against Gödel's P-deduction that if indeed, $[\vdash_P R] \Rightarrow \vdash_P g$, then P is inconsistent. Hence, there is no axiomatic system that does everything this proper Hilbertian formal system ought to—a semantical consideration in itself.

If, however, we agree to not read it squarely as given, then what Hilbert presents in "RULE ω*" above is not an inference rule at all. It is a preliminary sketch sufficiently ambiguous for multiple manifestations: one proving its own Incompleteness, in the other its inconsistency.

Feferman's Transfinite Progressions

And it appears that this is what was done (Feferman 1986); i.e., it has been decided that ω* will be read as $\omega_\downarrow$. And while it does not complete either P or P̄ as Hilbert had planned, $\omega_\downarrow$ does present a means of expressing a fundamentally semantic notion in surface-syntactic terms (See subsection "Lisbon & Pisa, 2014" above), while flattering the elder statesman with the denotation of an idea (i.e., Gödel's ω-consistency) that was never his. There is no "understanding" the ω-rule without accounting for this dynamic.

$$[\vdash_{Pi}R] \Rightarrow \vdash_{Pi+1} g$$

might also be understood as one of those manifestations (Schmerl 1982), as such deductive gaps might also be bridged formally, Feferman knew, in the Hilbertian spirit (at least up to Π_1-sentences), via a recursive hierarchy or "Transfinite Progression", of extensions to Gödel's system P (Feferman 1962) as devised 20 years earlier by Turing, wherein

(i) A_0 is given;

(ii) $A_{K+1} = A_K \cup \{Con_{A_K}\}$;

(iii) if κ is a limit number, $A_K = \cup_{i<\kappa} A_i$.

For proof of theorems higher up the arithmetical-order, Turing also considered the hierarchy of stronger extensions (A. Turing 1939)

(ii)" A_{K+1} consists of A_K together with all sentences of the form $(\forall x)\mathrm{Pr}_{A_K}(\phi(nm_x)) \rightarrow (\forall x)\phi(x)$.

but could come to no conclusion whether it worked. It is at this "point of departure", 20 years hence, that Feferman takes it up, taking care to flesh out in detail and with clarity the meta-reasoning behind Turing's ordinal-logic analysis of the "Gödel phenomena", which by this time had become a publishing industry stable.

The extending axiom that recurs above in (ii)" Feferman dubs a "Reflection Principle", explaining, "by which we understand a description of a procedure for adding to any set of axioms P certain new axioms whose validity follow from the validity of the axioms P and which formally express, within the language of P, evident consequences of the assumption that all the theorems of P are valid"—i.e., it is an amending axiom about the original axiom system,[62] and in Feferman's concern, about its soundness.[63] The resulting extended system itself is then said among logicians

---

[62]   Artemov & Beklemishev in *Provability Logic* (Beklemishev and Artemov 2004) put it, "Reflection is a general term describing an ability of a formal deduction system to formalize its own meta-reasoning. This normally includes internal representation of formulas, axioms, rules and derivations, semantics, etc., and ability to represent properties of those objects by formulas of the system."

[63]   However, as noted by a careful read, and as Gödel's Second Incompleteness theorem demands, Feferman's reflection does not *reflect upon itself*.

of a more linguistic bent[64] to be *Reflexive*—deductively reflexive,[65] not to be mistaken for *pure linguistic* "reflexivity" (discussed more fully in the coming section, "Warsaw, 1935", below.).[66]

In any case, Feferman finds that Turing's strengthened sequence yields in the end no stronger results, reaching no further than the Π_1-completeness of the original ordinal logics. Nor does assuming the Δ_n-soundness of P

$$\text{For any n, } [\ \vdash_P R(n) \Rightarrow R(n)\ ] \qquad \Delta_0$$

nor even, and to Feferman's concern, certain forms of the informal ω-rule

$$[\text{for any n, } \vdash_P R(n)] \vdash \forall x.R(x) \qquad \omega$$

of which, by means of Gödel's provability predicate, Feferman's reflection is an immediate formalization

$$\vdash_\omega [\ \forall x.\text{prov}_P(R(x)) \rightarrow \forall x.R(x)\ ].$$

Near the close of his original work, in connection with the informal proof of the 2nd Incompleteness Theorem, Gödel reflects (page [197]):

> ....all the concepts defined (or assertions proved) in Sections 2 and 4 are also expressible (or provable) [i.e., Capturable] in P. For we have employed throughout only the normal methods of definition and proof accepted in classical mathematics, as formalized in the system P.

perhaps anticipating Feferman's foundational concern. And earlier (page [187]):

> We now come to the goal of our elaborations. Let c be any class of formulae. We denote with Conseq(c) the smallest set of formulae that contains all formulae of c and all axioms and is closed under the relation "immediate consequence". c is called ω-consistent if there is no class-sign a such that:

---

[64]   Many at the time isolated behind the-then Iron Curtain.

[65]   Detlefsen (Detlefsen 1979) for example, defines:
   A theory T is said to be reflexive just in case for
   every finitely axiomizable sub-theory T_f of T, $\vdash_T$ Con(T_f)...
citing Mostowski (Mostowski 1952a). See also (Blanck 2017), (Shavrukov 1997), (Beklemishev and Artemov 2004), and (Kreisel and Levy, Reflection Principles and their Use for Establishing the Complexity of Axiomatic Systems 1968), via soundness $\Rightarrow$ consistency.

[66]   Though the similarly is perhaps not so close as their interdisciplinary authorships suggest.

(n) [Subst a(v, Z(n)) $\in$ Flg(c)] $\cap$ [Neg((v Gen a)) $\in$ Flg(c)]

so that, with c $\equiv$ P $\subseteq$ P*:

$\forall$x.prov(R(x)) $\vdash_{P*}$ ¬prov(¬g)	ω-consistency
$\vdash_{P*}$ [$\forall$x.prov(R(x)) $\rightarrow$ ¬prov(¬g)]	Deduction Theorem

from which equivalent supplementaries

$\forall$x.prov(R(x)) $\rightarrow$ g	Δ_0-soundness (or reflection)
g $\rightarrow$ ¬prov(¬g)	Σ_1-soundness

we may judge for ourselves whether the concern had not also there been *sufficiently* put to rest.[67]

Warsaw, 1935

the T-word

As mentioned earlier, in his introductory heuristic Gödel likens his Incompleteness Result to the semantic antinomies of the Liar and Richard's, which follow from the narrow-reflexivity (Lucy 1993) (Lyons 1977) common to natural languages, permitting statements about itself to be made within it, as when they refer to its regularities

前面的內容修改了後面的內容。

—a Chinese grammatical rule[68] rendered in Chinese Traditional script— or to the structure of its linguistic forms, as in

"dog" is monosyllabic

as well as to their truth conditions

"snow is white" is assertible

---

[67] $\vdash$¬g $\Rightarrow$ ¬[$\vdash$R] ω-consistency
      g $\Rightarrow$ [$\vdash$R]   V
      ¬[$\vdash$R] $\Rightarrow$ ¬g
      $\vdash$¬g $\Rightarrow$ ¬g Σ_1
      Thus,
           ω-consistency$\cap$Theorem V (Δ_0-completeness) $\Rightarrow$ Σ_1-soundness

[68]   Translated to English as, "What precedes modifies what follows."

—the mentioned form set off in quotes. And finally, fatally, this narrow-reflexivity may permit construction of statements that refer to themselves, such as the infamous Liar:

This sentence is false.

As clarified in section "Indirect Self-Reference" above (sub-section, "What does g not say?"), and admitted, if grudgingly, by the experts (Smith 2013) in the matter of this philosophic historiographical trifle that anyway lies outside his area of formal study, Gödel is simply mistaken. And for that brief moment in the introduction of the Incompleteness paper when it seems Gödel might overstep the boundary of his expertise, we hold our collective breath, then (out of respect) turn our heads the other way. In any case, to safeguard against genuine *semantic* antinomy inherent to such linguistic reflexivity, it is now convention that languages used in formal contexts be strictly non-narrowly-reflexive; that statements about one language *mentioned* in another be rendered in a distinct linguistic code, or script (Lucy 1993), these being object- and meta-languages, respectively. By this simple means, the Liar e.g., is rendered harmless, its anaphor redirected.

It is worth mentioning at this point, then, that up to now we have been somewhat lax in our grammar with respect to Feferman's work above. Our Δ_0-decidability and Σ_1-soundness statements e.g.,

$$[R] \Rightarrow [\vdash R] \quad \& \quad [\vdash R] \Rightarrow g$$

among others, are linguistically mixed, being neither P nor meta-P proper statements, ill-formed in this paradigm.

A Tarski-style, or nested, meta-language[69] speaks not only of the object-language form, or syntax, but also of its intended domain of discourse—(both syntax and intended dod then appearing separately in the meta-domain, itself likewise "nested" in a higher context[70]) sometimes

---

[69]   Among many other formulations: Benveniste's, for example (Benveniste 1971), in which the "formal language" does not itself include a discourse domain.

[70]   Hendricks' et. al. in *Knowledge Contributors* (Hendricks, Jorgensen and Pedersen 2013) put it that Tarski's "meta-level reasoning is first-order reasoning about the way statements are sorted into nested contexts..."

referring to one, sometimes to the other, sometimes referring to a rela-tion between the two.[71] It is a metalanguage specialized to the needs of the semanticist, if not particularly to those of the logician (Hendricks, Jorgensen and Pedersen 2013).[72] This dual-construction is augmented, making up the disjoint, by Tarski's Convention-T ("T" for "is true") or Equivalence Schema (ES), which states that for every translatably co-ref-erential pair (for which we use a notation, $\phi \sim \varphi$)

$$T[\phi] \Leftrightarrow \varphi \qquad (ES)$$

—proposed as a minimal or *adequacy* constraint, characterizing truth-predication,[73] an equivalence, if not definition, that any truth-predication must satisfy (Glanzberg, Truth/5.2 Minimalist Theories 2018).

Field (*Tarski's Theory of Truth* (Field 1972)) for example takes (ES) to follow from a coreferentiality

> In order to state the matter more generally, I introduce the term 'coreferential': two singular terms are coreferential if they denote the same thing; two predicative expressions are coreferential if they have the same extension, i.e., if they apply to the same things; and two functional expressions are coreferential if they are ful-filled by the same pairs. It is then easily seen that any departure from coreferenti-ality in translation will bring errors into T.

and warns

> It should be kept carefully in mind that the Quinean view that all we need do is clarify the term 'true', in the sense that this term is clarified by schema T (or by schema T plus a theory of translation to handle foreign languages; or by schema T plus the sort of thing alluded to in connection with Dummett), is not Tarski's view.

speaking, presumably, against Ramsey's stronger Equivalence-Thesis (Glanzberg, Truth/5.1 The Redundancy Theory 2018) reading of (ES) that

$$T[\phi] \text{ has the same meaning as } \varphi$$

taken up by many later-known deflationists, and here nicely explicated below in subsection "Lisbon & Pisa, 2014", by A. Yaqub (Yaqub 2008).

---

[71]   It is often said (unhelpfully) that a Tarski meta-language holds a *copy* of the object-language, for it is a copy *dismembered*, its syntax *mentioned*, functioning only by name, referring, not over an intended domain, but naming itself.

[72]   Also differing sharply from the ordered-metalanguage preferred by linguists (Lyons 1977) (Lucy 1993) (Benveniste 1971).

[73]   See Tarski's Semantic Conception of Truth (Tarski 1944) (esp. sections 3, 4, 5, and 9) for much of the following usage.

Similarly, Leitgeb, in *Truth as Translation* (Leitgeb 2001, 281-307) writes, "T. According to Tarski's Convention T, the adequacy of a truth definition is (implicitly) defined relatively to a translation mapping from the object language to the metalanguage," to which Raatikainen in *Truth, meaning, and translation* (Raatikainen 2008) adds:

> Tarski explicitly points out the difference here between his own approach and that of Carnap, according to which we regard 'the specification of *conditions* under which sentences of a language are *true* as an essential part of the description of this language.' (See, pp. 373, note 24; my emphasis). For Tarski, on the other hand, the interpreted object language is instead specified simply through its metalinguistic translation (see e.g., pp. 170–1; cf. Fernandez Moreno 1992, 1997; Milne 1997, Feferman 2004).

Tarski's Convention-T or (ES) originally appears as a "Translational adequacy" condition on T. Yet many refer to it as a "material adequacy". Schantz, for example. On page 4 of his introduction to *What is Truth?*, Schantz says (Schantz 2002), "...Tarski wanted to rehabilitate the concept of truth to a scientifically and metaphysically respectable concept....'materially adequate'... 'formally correct'....'Convention T'....'its translation'..."

Likewise, we find in M. Glanzberg's Stanford-Encyclopedia article on "Truth" (Glanzberg, Truth/5.2 Minimalist Theories 2018):

> The equivalence principle looks familiar: it has something like the form of the *Tarski biconditionals* discussed in section 2.2. However, it is a stronger principle, which identifies the two sides of the biconditional—either their meanings or the speech acts performed with them. The Tarski biconditionals themselves are simply material biconditionals.
> A number of deflationary theories look to the Tarski biconditionals rather than the full equivalence principle. Their key idea is that even if we do not insist on redundancy, we may still hold the following theses:
> 1. For a given language L and every φ in L, the biconditionals ⌜⌜φ⌝ is true if and only if φ⌝ hold by definition (or analytically, or trivially, or by stipulation ...).
> 2. This is all there is to say about the concept of truth. We will refer to views which adopt these as *minimalist*. Officially, this is the name of the view of Horwich (1990), but we will apply it somewhat more widely.

And maybe it doesn't matter what one calls it, how one characterizes it, so long as one is clear about the objective facts of the matter. When we say (as below) that "truth is translation", we do not mean as Davidson apparently does that the T-sentences are adequate to a "meaning theory"; no. The analogy we draw with Quine's "truth is disquotation" should make it clear enough. It is only, again, that the condition on T

$$(\phi \sim \varphi) \Rightarrow (T(\phi) \Leftrightarrow \varphi)$$

we, like Leitgeb cited above, call, characterize as a "translational" rather than "material" adequacy.

And as in reflexive metalanguages, it is sometimes said that "truth is disquotation" (Field 1994),

> 'Snow is white' is true if and only if snow is white

we may here venture the mnemonic that *truth is translation* (Leitgeb 2001) (Raatikainen 2008),

> 'Snow is white' est vrai si, et seulement si, la neige est blanche

the English source translating to a French target.

We immediately set right then our earlier oversight

$$[R] \Rightarrow [\vdash R] \ \& \ [\vdash R] \Rightarrow g$$

by direct application of Conv-T (ES)

$$[R] \Rightarrow [\vdash \mathbf{R}] \ \& \ [\vdash \mathbf{R}] \Rightarrow g$$

where R is the meta-P translation of $\mathbf{R}$, satisfying Conv-T, g being that of $\mathbf{g}$.

It is only in the background, then, that the highly provocative T-word shows up. In our calculations and deductions here, it does not show up at all, any more than it does or needs to in the claim

> '2+2=4' is provable if and only if 2+2=4.

the truth predication of $2+2=4$ in meta-P being perfectly redundant to meta-P's own, 2+2 =4.

While this translational account of truth is perfectly sufficient to our minimal needs here, whether "this is all there is to say about the concept true" (Glanzberg, Truth/5.2 Minimalist Theories 2018) or whether there might be more, is a discussion we leave for the coming sections.

For the moment, we only observe that if not his full account of truth, this is certainly one to which Tarski was comfortable retreating when pressed. On page 103 of *Two Types of Deflationism* (Yaqub 2008) Yaqub cites:

I would like to add that this insight goes back to Tarski. In 'The Semantic Conception of Truth" (Tarski 1944), he defends his conception against the charge that it is "involved in a most uncritical realism" (p. 33). He writes:

> For these words [such as 'in fact'] convey the impression that the semantic concep-
> tion of truth is intended to establish the conditions under which we are warranted
> in asserting any given sentence, and in particular any empirical sentence. However,
> a moment's reflection shows that this impression is merely an illusion; and I think
> that the author of the objection falls victim to the illusion which he himself created.
> In fact, the semantic definition of truth implies nothing regarding the conditions
> under which a sentence like:
>> (1) snow is white
> can be asserted. It implies only that, whenever we assert or reject the sentence,
> we must be ready to assert or reject the correlated sentence:
>> (2) the sentence "snow is white" is true.
> Thus, we may accept the semantic conception of truth without giving up any epis-
> temological attitude we may have had; we may remain naive realists, critical real-
> ists or idealists, empiricists or metaphysicians—whatever we were before. The se-
> mantic conception is completely neutral toward all of these issues, (pp. 33-34)

And so we have here via Tarski's (ES) two sentences that assert the same thing, have the same sense and reference; it is not a condition under which one is satisfied and we then say that it has some property T. No, in Tarski's semantic conception (or theory) "semantic" is taken in its most general sense: "The branch of linguistics or philosophy concerned with meaning in language" (ODE 2022).

Similarly, Raatikainen in *Truth, Meaning, and Translation* (Raatikainen 2008) continues at length:

> In accordance, Tarski described the metalanguage in the following ways:
>> '...the metalanguage contains both an individual name and a *translation* of
>> every expression (and in particular of every sentence) of the language stud-
>> ied...'(Tarski 1935, p. 172; my italics)
> *and*
>> '...to every sentence of the language . . . there corresponds in the metalan-
>> guage not only a name of this sentence of the structural-descriptive kind, but
>> also a sentence having the *same meaning*...' (Tarski 1935, p. 187; my italics).
> However, one could point out that Tarski's approach still assumes the notion of
> *meaning,* in the disguise of translation, or the sameness of meaning... He explained
> his understanding of 'semantical concept' as follows:
>> 'A characteristic feature of the semantical concepts is that they give expres-
>> sion to certain relations between the expressions of language and the objects
>> about which these expressions speak, or that by means of such relations they

characterize certain classes of expressions or other objects.' (Tarski 1935, p. 252)

In contrast, I submit that it is possible to view translation, in this context, as a purely syntactic, effective mapping between two languages, without assuming any relations between either language and objects about which they speak. Translation, so viewed, is *not* a semantical concept in Tarski's sense, and does not presuppose truth or related notions (most importantly, satisfaction, by means of which the others can be defined). Hence, it seems to be, after all, admissible for Tarski to presuppose such a notion of translation in his approach without begging the question (cf. Milne 1997; see also below).

The two oft-cited *difficult cases* for this translational account—blind truth-ascriptions (Schindler and Picollo 2019) in non-autonomous (Lucy 1993) or homophonic (Leitgeb 2001) metalanguages, and *extended* object-language generalizations—are also found to in the end give way to redundancy.

Blind-ascriptions to non-transparent terms (Schindler and Picollo 2019), or "indirect" (Shapiro 1998), "lazy" endorsement (Grover, Camp and Belnap 1975), such as

Newton's Second Law is true.

common to natural language, might always be taken as prosententially anaphoric (Grover, Camp and Belnap 1975) (Shapiro 1998),[74] thus non-reflexive and redundantly superfluous to their direct counterparts

The net force on an object equals its mass times the resulting acceleration.

however unwieldy, inconvenient.

The same, *often*, with regard to generalized endorsements such as

Everything the government says is true.

Here, in the context of Gödel's quadruple linguistic analysis, we have at our disposal the additional P/P' translation

$$T(R) \Leftrightarrow T(G(R)) \qquad \text{meta-P/meta-P'}$$

by means of which Gödel's 46th Capture

---

[74]   For a review of the Prosentential-Disquotational-Minimalist distinctions, see e.g., Michael Williams *Critics of Deflationism* (Williams 2002) In Schantz's *What is Truth*, or Horsten's *The Tarskian Turn* (Horsten 2011).

$$\vdash R \Leftrightarrow T[prov(G(R))] \qquad \text{meta-P}$$

helps facilitate demonstration of the redundancy. Thus it happens that Gödel's Theorem V

$$\forall x.(R(x) \rightarrow prov(G(R(x)))) \qquad P$$

is read by the ordinary natural-language meta-mathematician

$$T(R(n)) \Rightarrow \vdash R(n)$$

i.e.,

"P proves all true primitive-recursion relations"

without Gödel ever having used the T-word.[75]
We repeat a similar disquotational exercise for Theorem V's *soundness* converse in our next "Ohio, 1998" section (the paragraph, "Offended by...") for the specific case of S. Shapiro's "thick" truth predication there at issue.

As for Feferman's usage, neither his "reflection" nor the "reflexivity" of the extended deductive system in which it appears coincides with or corresponds to that of Lucy and the linguists—a reflexivity of long-standing and in wider use,[76] with its capacity for semantic paradox.

Indeed, Artemov & Beklemishev in *Provability Logic* (Beklemishev and Artemov 2004, 229-403) even broaden Feferman's reflection to

> ...a general term describing an ability of a formal deduction system to formalize its own meta-reasoning. This normally includes internal representation of formulas, axioms, rules and derivations, semantics, etc., and ability to represent properties of those objects by formulas of the system...

which one would understand were theories whose theorems exhibited this *reflectivity* also said to be *reflective*. Instead, we are offered a usage in which they are said to be *reflexive*.

Detlefsen in *On Interpreting Gödel's Second Theorem* (Detlefsen 1979), e.g., defines

---

[75] I.e.,

	P		Meta-P (Δ_0-completeness)	Meta-P (capture)		P
R	$\Leftrightarrow$	T(R)	$\Rightarrow$	$\vdash R$ $\Leftrightarrow$ T[prov(G(R))]	$\Leftrightarrow$	prov(G(R))

[76] The reflexivity of semantically closed languages, in Tarski's usage.

A theory T is said to be reflexive just in case for every finitely
axiomizable sub-theory

$$T_f \text{ of } T,\ \vdash_T \text{Con}(T_f)...$$

citing Mostowski's *On Models of Axiomatic Systems* (Mostowski 1952a)
(See also (Detlefsen 1979), (Blanck 2017), (Shavrukov 1997), (Beklemishev and Artemov 2004), and (Kreisel and Levy, Reflection Principles and
their Use for Establishing the Complexity of Axiomatic Systems 1968), via
soundness $\Rightarrow$ consistency.), in which the term does not appear. Blanck
(*Contributions to the Meta-mathematics of Arithmetic* (Blanck 2017))
makes the same claim, and also cites Shavrukov's *Interpreting reflexive
theories in finitely many axioms* (Shavrukov 1997) where the usage does
indeed appear.

What might be the rationale behind this usage? Some of Gödel's
early 46 Captures have syntactic consequence, and to the extent that one
may say they "indirectly refer" to the linguistic grammar of P (e.g., VARI-
ABLE and FORMULA), one might also indulge the mnemonic that they are
properly, then, linguistically reflexive. Perhaps "indirectly reflexive". If
one ventured further to include with linguistic syntax the purely syntactic
inference rules of P expressed in the language as "deductive grammar"
(in the manner of Carnap, who includes it in the Logical Syntax (Carnap
1934) (Bar-Hillel 1954) of an inferential language), then Gödel's deductive
Captures too (e.g., AXIOM and PROVABLE), along with Feferman's inferential reflection, might also be included. Whether Feferman arrives at this
usage by such tortured convolution, we are nowhere told. In any case,
the usage is not without dangers.

One wonders then at the wisdom of Feferman's usage, that seems
only to play into the misconception previously mentioned, and really understood by all: Gödel's original, if pardonable philosophic overstate-
ment, as well as Hofstadter's and a thousand other's false-friend misconceptions of the Gödel sentence. No, Feferman's reflection says nothing of
itself or co-axioms, and the unfortunate usage does not advance the
much needed *clear and distinct Idea* (Locke 1975) of Gödelian Incompleteness.[77]

---

[77]   While Tarski's Conv-T (e.g., in French-English) are indeed generally *metalinguistically*
reflexive (e.g., in French-English), its *Prosentential* reading (e.g., in English-English) is

Ohio, 1998

Newark vs. Columbus

"Worth mentioning" at this point, we were just saying, as the careless-ness pointed out there and duly corrected—the mixing of propositional assertions with meta-assertions not uncommon since the writings of David Hofstadter—has in the past 100 years been more unhelpful and det-rimental to the setting forth of *clear and distinct Ideas* of mathematical fact (Locke 1975)—i.e., of *fact*—than anyone since Gödel might have feared. For there are truths, and then there is the Truth, an actuality that transcends ordinary fact—or so there had been as long as anybody could remember.[78] Then came the turn of the 20th century and Gottlob Frege, who followed by others would reduce this talk of "truth" to a notational redundancy of linguistic use, to such a degree that to say, "I smell the scent of violets" *is true*, would be no more than to say "I smell the scent of violets" (Frege 1918).

Unhappy with this turn, though powerless to do much about it, True believers kept quiet, content for the moment to be left alone. Then in the early 1930s there was news from Königsberg of the discovery of a *true* yet unprovable mathematical statement. From Warsaw soon after came word of a companion proof of the *Undefinability of Truth* (Tarski, *The Concept of Truth in Formalized Language* 1933) itself upon the most "far-reaching revision in the theory of truth..." (Davidson 1967)[79] in a millen-nia.[80]

---

[78] not. And neither, again, strictly, properly speaking, are any of Gödel's correspond-ences except, we might mention, in Davidson's unconventional material sense of meaning (Davidson 1967)

[78] Say, since Plato.

[79] Or better, from Soames' *Understanding Truth* (Soames 1999): "the far-reaching dif-ference between Tarski's truth predicate and our ordinary notion of truth..."and Can-tini's *Truth in the 20th Century* (Cantini 2009): "The importance of Tarski's work is dif-ficult to overestimate. He immediately realizes that (p. 401) his own semantical meth-ods prompt him a way to show that Gödel sentences are true and yet formally unde-cidable."

[80] Whether Gödel himself took the results very seriously is doubtful. He certainly never mentions them in print (Krajewski 2004). Consider also Murawski's *Tarski vs Godel* (Murawski 1998): "It [the statement of Tarski's theorem] was followed by a descrip-tion of the idea of the proof and then by a sketch of the proof. The theorem was

This pleasant coincidence *forced* Truth back into the mouths of philosophers and onto the pages of their research journals (Kohler and
Wolenski 2013), only now with a renewed vigor of mathematical rigor
and respectability (See e.g. page 1 of Wolenski's (Kohler and Wolenski
2013)). T-Transcendentalists, let us call them[81] (this "transcendental"
truth predicate eluding formalization (Azzouni, Tracking Reason: Proof,
Consequences, and Truth 2006)), quickly found their feet and began to
speak again with confidence of truth in transcendence—a transcendence[82] over "verification", i.e., proof—that adherents to the Frege line,
"deflationists" as they would eventually be called, could not ignore.[83] Nor
did they. Through the post-Gödelian war years (Ayer, Carnap, Wittgenstein), Postmodernism (Quine, Field, Horwich, Davidson,...?) to now, the
core deflationist line runs firm (Stoljar and Damnjanovic 2014):

- the reference to truth never adds anything to the sense
- the words true and false are not used to stand for anything
- truth and falsehood are not genuine concepts
- there is no genuine truth predicate
- etc.

T-transcendentalists too (Putnam, Dummett,...?) held to their guns and
more.[84] For soon, and in a truly extraordinary turning of tables, they
would abandon what had become their usual defensive posture and set
themselves on a vigorous offensive. By the 1990s they had devised an
ingenious argument designed to secure the metaphysical realism inher-

---

proved by diagonalization, hence it is closely connected with methods and results de-
veloped by Kurt Gödel in the paper *Über formal unentscheidbare Sätze der 'Principia
Mathematica' und verwandter Systeme. I (1931)...* "

[81]   In the spirit of Shapiro's characterization of Palinor (Walsh's *Knowledge of Angels*
(Walsh 2011)) as godless proto-deflationist (Horrigan 2014).

[82]   As Field in *Tarski's Theory of Truth* relates (Field 1972), "As a result of Tarski's teaching, I no longer hesitate to speak of 'truth' and 'falsity'," wrote Popper'; and Popper's
reaction was widely shared.²

[83]   Who in due course would be known as "deflationists", but whose core belief remained no more than the simple redundancy of lexical truth. Trained as mathematicians, most were not always sufficiently disposed to answer. In any case, and for different reasons, logicians and linguists alike have seized upon the term.

[84]   Perhaps emboldened then (and a little surprised) by the relative ease of their redis-
covered freedom, and the relative inexperience if not naiveté of mathematicians of
the day in matters of Truth.

ent to their notion of truth, breaking the stalemate while calling deflationism out on its own terms: If the deflationist, with all others in the civilized world, believes in the truth of the Gödel sentence while simultaneously in the redundancy of "truth", then he or she must prove this truth without use of the word "truth", except perhaps at the very end of the proof, disquotationally.[85] For only such a proof, "thin" in respect of the "substantiality" of the "truth" concept, the argument continues, is available to the honest deflationist; only a deductive system to whose calculus Truth predication adds nothing—a system "conservative over T"—is deflationarily legitimate, i.e., "licit".[86]

Details of the new deflationary constraint inherent to this demand are set forth in 28 single-spaced and column pages of the October 1998 issue of *The Journal of Philosophy*, under the authorship of Ohio State Newark's eminent philosopher Stewart Shapiro (Shapiro, Proof and Truth: Through Thick and Thin 1998) who singles out Michael Dummett's *semantical argument* for the truth of the Gödel sentence as, "the most sensitive to the actual [Sic] logical structure of the argument" to date in all of 80+ years (Tennant 2002):

> G is a universally quantified sentence (as it happens, one of Goldbach type, that is, a universal quantification of a primitive-recursive predicate). Every numerical instance of that predicate is provable in the system S. (This claim requires a subargument exploiting Gödel-numbering and the representability in S of recursive properties.) proof in S guarantees *truth*. Hence every numerical instance of G is *true*. So, since G is simply the universal quantification over those numerical instances, it too must be *true*.

Shapiro continues:

> Michael Dummett comes to a similar conclusion. Suppose that someone advances the deflationist "view that we can have no notion of truth for arithmetical statements... beyond a strictly disquotational one". Dummett argues that on this view, "an intuitively correct formal system will be one... whose axioms we *treat* as true, but no reflection will guarantee that we have any reason to think it consistent, nor, hence, any to treat the undecidable statement as true." In other words, the strictly disquotational notion of truth does not allow the inclusion of formulas with the truth predicate in the induction scheme, and does not support the relevant generalization (*). If the consequence relation is first-order, then the notion of truth (for

---

[85]   "...without explicit use or mention [sic] of the truth predicate" (Tennant 2002),... and preferably without mention of other T-words, such as Tarski.

[86]   More specifically, a theory is said to be conservative over truth when adding its predicate proves no more than the original theory.

L) in A" is more substantial than the mere disquotational one, and so this notion is not available to the deflationist.

This "relevant generalization" being no more than that "all P theorems are true"

$$(*) \quad \forall x (\exists y PRF(x,y) \rightarrow T(x))$$

i.e., a non-disquotational soundness of P. This, alongside Schindler's (Schindler and Picollo 2019) "All theorems of Arithmetic are True" example, who goes on to quote Quine

> We may affirm the single sentence by just uttering it, unaided by quotation or by the truth predicate; but if we want to affirm some infinite lot of sentences that we can demarcate only by talking about the sentences, then the truth predicate has its use. (Quine [27, p. 12])

and Cantini (Cantini 2009), also quoting Quine (from *Philosophy of Logic* (W. Quine 1970)), "This is the reason why the notion of truth is so precious to us: it is one of the means by which finite minds are able to apprehend the infinite."

As it happens, Gödel's Theorem V, too

$$\{\text{for any } \mathbf{n}, [T(\varphi(\mathbf{n})) \Rightarrow \vdash \varphi(\mathbf{n})]\} \text{ for any PRR } \varphi$$

is of this type. Such generalizations, presumably, defy deflation requiring our deflationist to perform nonconstructively infinite acts of substitution.[87]

---

[87]   For a sample of this thinking, see e.g. Halbach's *Disquotational Truth & Analyticity* (Halbach 2001), and his Stanford Philosophy article (Leigh and Halbach 2020) on "generalizations" as non-constructably "infinite conjunctions". See also Stoljar's Stanford Philosophy article on deflationary truth (Stoljar and Damnjanovic 2014), first paragraph.
See also Schantz's, *What is Truth?* (Schantz 2002), page 7 of Introduction:
The utility of the truth predicate….infinite conjunctions….disquotation…
And again from Stoljar's *Deflationary Theory of Truth* Stanford Philosophy article (Stoljar and Damnjanovic 2014):
On the contrary, however, advocates of the deflationary theory (particularly those influenced by Ramsey) are at pains to point out that anyone who has the concept of truth in this sense is in possession of a very useful concept indeed; in particular, anyone who has this concept is in a position to form generalizations that would otherwise require logical devices of infinite conjunction….
Given deflationists place such heavy emphasis on the role of the concept of truth in expressing generalizations, it is ironic that some versions of deflationism have been criticized for being constitutionally incapable of accounting for generalizations

Thick truth predication via Dummett's argument, *de rigueur* to ordinary grammar (as in (*)) Shapiro concludes, *is therefore necessary* to the establishment of the truth of the Gödel sentence:

$$[\vdash R]$$
$$\forall x.(\ prov(x) \rightarrow T(x)) \qquad (*)$$
$$(\vdash(\phi) \Rightarrow T(\phi)\ ), \text{ for any } \phi$$
$$[T(R)]$$
$$T(g)$$

whatever deflationist protests to the contrary,[88] and Shapiro's deflationist-of-straw looks on speechless, stunned.

But "why not claim on the contrary that truth must be a substantial property because universal generalizations about truth are not deducible from MT?" (David 2002, 174), and the whole of Shapiro's argument—at least in respect of the Gödel sentence insinuated here—might be seen to rest upon this fancy alone. This is what Marian David queries in his chapter *Minimalism and the Facts about Truth* (David 2002) of Schantz's *What is Truth?* (page 174). Although Tarski's Convention-T (David's propositional "(E)") comprising David's "MT" there is itself such a "universal generalization about truth" (*the* generalization, according to deflationists)— his "IMPx" of (7.1) being the metalinguistic translation of "x"— David, Shapiro and followers consistently leave out of account the fact that the P-soundness assumption is a statement about P-statements, in light of the restriction that a formal language may not speak of its own statements (that must happen in its meta-language where the deflationary-sanctioned Conv-T freely does its work). As for

Everything the government says is true.   (NL)

and like perfectly grammatical natural-language generalizations, it is worth reminding ourselves of the semantic-paradox-solving origins of the

---

about *truth* (Gupta 1993) (Halbach 1999) (Soames 1999) (Armour-Garb 2004). For example, theories that implicitly define truth using only the instances of (ES-prop) do not allow us to derive a generalization like (Conjunction).

[88]   I.e., the assumption being that the truth of the Gödel sentence, as parcel to Incompleteness, may not be established without assuming that P is sound (which *as a generalization* cannot be expressed without use of the truth predicate) and so is itself then thick with truth.

modern-day formal-language (Belnap 2006). There, (NL) is simply un-translatable to the formal syntax; such generalizations simply, thankfully, cannot show up in the logico-mathematical business.

Offended by the unfairness of Shapiro's one-sided and, frankly, wordy proceedings, while "holding no brief" for deflationism itself,[89] Ohio State Columbus' Neil Tennant (Tennant, Deflationism and the Godel Phenomena 2002) takes up Shapiro's challenge to derive the truth of g under the proposed constraint that "...no notion of truth beyond the disquotational one" be applied deflationarily, bringing it to bare on the schema corresponding to Shapiro's (*).

Axiom (*) of course *P-Captures* the Meta-P' soundness condition[90]

$$\vdash_{P'} x \Rightarrow T'(x), \quad \text{for any } x \quad \text{meta-P'}$$

while ignoring the meta-P/P' Correspondence (See above "Correspondence" subsection)

$$\begin{array}{ccc} \text{meta-P'} & & \text{meta-P} \\ T'(x) & \Leftrightarrow & T(G^{-1}(x)) \end{array}$$

afforded by their common Conv-T ordinary-language translation templates always present in Tarskian metalanguages,[91] which makes available the twin soundness condition

$$\vdash_{P'} x \Rightarrow T(G^{-1}[x]), \quad \text{for any } x$$

whose disquotation into an extended $\tilde{P}$-Capture for primitive recursive relations R

$$R(a) = G^{-1}[x], \text{ for some } a$$

readily proceeds via Gödel's 46 provability Capture to the axiom schema

$$\forall y[prov(G[R(y)]) \rightarrow R(y)] \qquad \tilde{P}$$

---

[89]   Acting instead as *amicus curiae* (Tennant 2010).
[90]   Cf. disquotation for Gödel's Theorem V, beginning paragraph "Here, in the context..." in previous *Warsaw, 1935* subsection.
[91]   I.e.,

$$\begin{array}{ccccc} & \text{meta-P'} & \text{meta-P} & & \\ T'(x) & \Leftrightarrow & \varphi & \Leftrightarrow & T(G^{-1}(x)) \end{array} \qquad \text{for some ordinary-language } \varphi.$$

See subsection *Correspondence* above, and paragraph, "A Tarski-style..", in previous subsection *Warsaw, 1935*.

or in the shorthand we have agreed to, simply

$$\forall y[prov(R(y)) \to R(y)]$$

for any PRR (i.e., Δ_0), R.

Shapiro's thick reflection then

$$\forall x[prov(x) \to T(x)] \quad (^*) \qquad \check{P}$$

gives immediate way to Tennant's thin Uniform Reflection schema[92]

$$\forall x.(Prov_{PA}(\phi(x)) \to \phi(x)) \qquad UR_{p.r.} \qquad \tilde{P}$$

and corresponding argument for the truth of g:

$$[\vdash R]$$
$$\forall x.(Prov_{PA}(\phi(x)) \to \phi(x)) \qquad UR_{p.r.}$$
$$[\vdash(R(n)) \Rightarrow R(n)], \text{ for any } n$$
$$[R]$$
$$g$$
$$T(g)$$

So came and passed quickly into the logico-mathematical annals this second in the yet unfolding sequence of dramas around the struggle for the final word in Gödelian truth.[93] And there it lay dormant[94] till a decade later it came to the attention of post-millennial graduates of the school of professional philosophy, bored from years of idle speculation. Feeling

---

[92] Which is nothing more than Gödel's own Theorem V in converse (See paragraph, "Gödel begins down…" in subsection *Konigsberg, 1930*, above.), and perfectly suited to the needs of Incompleteness with quotation for T(g).

We only caution a couple details worth clarifying, if inconsequential to Tennant's final application of this result: The disquotation resulting in $UR_{p.r.}$ is not likened to the anaphoric reference of Tennant's preferred deflationary Prosentential Theory of Truth (Grover, Camp and Belnap 1975). See e.g., '(pa)' of Tennant's section '8.3', while in fact, $G[\phi(x)]$ and $\phi(x)$ are rendered in distinct linguistic codes (cf. footnotes (71) and (73), above), and that while the Σ_1-soundness of $UR_{p.r.}$ is sufficient to prove Undecidability, Tennant's *admittedly weaker* "Soundness extension of S" restricted to Π_1 formulas would seem closer to the "lightest hammer" for Incompleteness (and so for the truth of g) that he seeks.

[93] References, (Ketland 1999) (Azzouni 1999) (Ketland, Conservativeness and translation-dependent T-schemes 2000) (Halbach 2001) (Field 1999) (Gauker 2001) (Sandu and Hyttinen 2004) (Ketland 2005) (Tennant 2005).

[94] Excepting a brief flurry in 2010 (e.g., (Cieśliński 2010) (Ketland 2010) (Tennant 2010), Tennant again having the last word.

this matter of "truth" to be within their purview, as perhaps it is, they weigh heavily in.

Lisbon & Pisa, 2014

> Philosophers have long been aware of the philosophical objections against the deep theories of truth…. Defenses have been mounted against these charges, and these defenses have subsequently been critically examined. Thus, a lively debate concerning the nature of truth has been carried on for centuries, and a merry time was had by all.
>
> Horsten, *The Tarskian Turn*

Gödel of course proves Undecidability while never formally laying claim to Incompleteness—perhaps for the reasons we suggest, perhaps leaving Incompleteness as a sweetener for prospective adherents. Something of the latter seems to be the line taken by the next to join the fray, a company of continental philosophers entering on the side of deflationism (Piazza and Pulcini 2015), though it is the deflationist strawman who again takes the brunt. Neither does Tennent escape unscathed; while his defense of deflationism against unfair treatment is noble enough, he goes about it wrongly, misrepresenting the very deflationism he claims to defend (ibid). The article proceeds early on through a list of like criticisms, bold pronouncements and loosely associated philosophic notions, and it is difficult to keep pace:

1) Genuine Deflationism takes no particular stance against the "The Substantialist Dogma" that arithmetic is non-conservative over truth-predication.

2) Neither ought we to associate deflationism with a reading of truth, thick or thin, that endorses ad-hoc reflection rules, like Tennant's $UR_{p.r.}$, which are no more than the old anti-deflationary soundness in clever Tarskian disguise.[95]

3) Genuine Deflationism may be summarized by the dictum, truth=proof.

4) The notion of a deflationary *semantical* derivation of GI is incoherent.

---

[95]   That like Achilles, however, remaining forever, and at every level, one deductive hole behind Zeno's tortoise (§ 5.2.3).

5) Far from its Incompleteness, Gödel proves the "inexhaustible deducibility" of Peano Arithmetic.
6) The unrestricted ω-rule is not only non-constructive, but also semantic.
7) Etc.

Three-quarters the way in, and we come to the crux, that notwithstanding the deflationist fine insight into the redundancy of linguistic truth, his "ontological commitment to natural number" remains a conventional fatal flaw implicit to his acceptance of ordinary (i.e., unrestricted) Universal Generalization—an extensional *quantification* over the "all" of an infinite domain,

$$[R] \Rightarrow g$$

that "enjoins us to assume a numerical ontology..." Their solution and claim to novelty in the crowded market for deriving the truth of the Gödel sentence is a new rule of inference

R(n) derived in PA for *any arbitrary* n, derives *g* in the extended system
$$S^* \qquad \omega_\downarrow$$

to replace Tennant's reflection

$$\vdash_{S^*} [\forall x.\text{prov}(R(x)) \rightarrow g] \qquad\qquad UR_{p.r.}$$

For while Tennant's supposed "thin" proof for the truth of the Gödel sentence is, they continue, no more than a *thinly* veiled semantic, thus thick proof, $\omega_\downarrow$ is "purely syntactic"—the notational subtext being that as Gödel makes syntactical his hasty semantical introductory heuristic by renouncing contentual Σ_1-soundness in favor of non-contentual (thus *deductively weaker*) ω-consistency, they do likewise here with $\omega_\downarrow$ in respect of $UR_{p.r.}$.[96]

Notational suggestions aside, the relative comparison of these differences separated as they are by a century of like novel inferences might be a stretch, and on close examination of the pair before us ("S*-proof of

---

[96]   Though again, the actual and well known "purely syntactical" proof of the disquotational truth of the Gödel sentence *g* is straightforward enough, involving neither Gödel's own ω-consistency, $\omega_\downarrow$ here, reflection rules (including $UR_{p.r.}$), nor unapologetic Σ_1-soundness, none of these. This, while the Semantical Derivation remains, yes, semantical...

G" on page 577 of reference (Tennant 2002); the "prototype juncture" on page 81 of (Piazza and Pulcini 2015)) it is hard to ignore the impression that one is witnessing one and the same proof, as indeed they are. For while

$$\Delta_0\text{-soundness} \cap \Sigma_1\text{-soundness} \Rightarrow \omega\text{-consistency}$$

and

$$\omega\text{-consistency} \Rightarrow \Sigma_1\text{-soundness}$$

though

$$\neg(\ \omega\text{-consistency} \Rightarrow \Pi_1\text{-soundness}\ )$$

in which case Gödel's ω-consistency is purely formal and weaker than the $\Sigma_1 \cap \Pi_1$ needed to semantically derive Undecidability, $UR_{p.r.}$ & $\omega_\downarrow$

$\vdash_{S*} [\forall x.prov_P(R(x)) \to g\ ]$ $\qquad\qquad$ $UR_{p.r.}$

$T[\forall x.prov_p(R(x))\ ] \vdash_{S*} g$ $\qquad\qquad$ Deduction Theorem

$prov(R(n))$, for any arbitrary n $\vdash_{PA} \forall x.prov(R(x))$ $\qquad$ UG

∴ $\qquad$ R(n) derived in PA for *any arbitrary* n, derives *g* in the extended system S* $\qquad$ $\omega_\downarrow$

are simply two equally contentual sides of the same coin, equivalent to Δ_0-soundness, hence insufficient to derive Undecidability.[97] As regards the ontological commitments of their $\omega_\downarrow$, our philosophers point to its eminently intensional "constructivisation"—rather, that of its premise, "$\forall x.prov(R(x))$", and one has to wonder, who is being implicated here? Who claims to establish this Π_1 wff by any other than constructive means? Tennant? [98] Perhaps Gödel? [99]

In any event, the point of philosophic refinement by which (Piazza and Pulcini 2015) intends to strengthen the non-contentuality claim of their case, is that while an extensional reading of the universal *quantifier*

---

[97]   Which if nothing else brings to mind Ketland's reply to Tennant (Ketland 2005) reviewed in section "4.2 Ketland's reply: reflection principles are truth-theoretically justified" of Vidal-Rosset's *Does Gödel's Incompleteness Theorem Prove That Truth Transcends Proof?* (Vidal-Rosset 2006); i.e., that behind every "purely-syntactic" rule of inference is a dirty epistemic secret. Such reflection rules are also of course "formal" in the ordinary inferential sense, and what is clear from context is that Gödel's claims about his own ω-consistency refers to its "non-contentuality".

[98]   See page 577, under 'S*-proof of G' of Tennant's (Tennant 2002).

[99]   See diagram (B) in subsection *Konigsberg, 1930*, above.

"∀x" (i.e., as "for *all* x", e.g., as an infinite conjunction) amounts to an ontological commitment to the existence of numbers, being then *in some sense* semantical, contentual, a restricted intensional reading, "for any arbitrary x", does not, and as a *qualifier* goes rather to *define* what numbers are. The Deflationarily acceptable, "licit", non-contentual premise "∀x.prov(R(x))" (whether for ω, ω₁, or UR_pr) must therefore be established or constructed, in the words of their much cited Cellucci (Cellucci 2009), "by a generic argument", or in those of Herbrand (Herbrand 1932), also heavily cited, "uniformly", Longo (Longo 2011), "prototypically", Bundy et al (Bundy, et al. 2005), "schematically",[100] Jamnik (Jamnik 2001), "diagrammatically", etc.

And all demonstrably correct, though of no novelty. Whether "∀" is read intensionally or extensionally, as "for any" or "for all", or if with Russell we take one reading to follow from the other,[101] or with Zardini (Zardini 2015) and Longo (Longo 2011) cited here, we do not; i.e., however one takes "∀" in its semantic clause (Shin 2015), there is but one means by which to infer, construct or establish this universal generalization: by the means Gödel employs on page 189 of (Gödel 1962), Tennant on page 557 of (Tennant 2002), our philosophers here on page 81 of (Piazza and Pulcini 2015). Each does so by the means set forth in the ordinary inference rule for Universal Generalization variously expressed

$$\frac{A(x)}{\forall x.A(x)} \quad , \quad \frac{C \supset A(x)}{C \supset \forall x.A(x)} \quad , \quad A(x) \vdash^x \forall x.A(x)$$

though uniformly applied, "A(x) for *any arbitrary* x".[102] Whether the above consequent *ought to then* be read "for all x" (as a "potential" infinite ever unfolding into an "actual" one), as Russell famously thought and

---

[100] Or indeed, by Shapiro's "indefinitely extensible" process, which like the rest harps on the qualitative definiteness of classifications ("classes"), and quantitative definiteness of a an entity mathematically set ("sets") that their ordinary-language etymologies, if not their current use in the field, invites.

[101] From "free" to "bound" variables, or Frege's Greek to Latin alphabet, whereas Kleene explicitly demonstrates the grammatical necessity of the distinct though "*synonymous*" notations (page 151 of (Kleene 1971)), a synonymy between terms in distinct languages, meta and object.

[102] In line, we think, with Locke's conception of generality as an Idea, via *succession*, part and parcel of that of number ((Locke 1975), Chapter XVII, section 9) and not one to

formulated in terms of free to bound variables (B. Russell 1988), or as strict "synonymy" as Kleene and apparently Frege took it (see e.g. page 66 of his (Frege 1967) and §11 Generality of (Frege 1879))—the rule then being no more than translational inference[103]—is a question that does not concern us here.

Finally, there is the true, the transcendental Truth, and now the "literally true", to put it rather tiresomely, as does A. Yaqub in an otherwise

---

be applied as though external to it (and for which the imagined 'problem of generality' (Cellucci 2009) disappears). Tait ((Tait 1981) section II), five centuries later and in the language of finite *sequences*, echoes much the same:

f: ∀xF(x) should mean fa :F(a) for arbitrary a: A

given primitive-recursive algorithm f, which in different terms is nothing more than what is now the standard rule for Intuitionistic (Iemhoff 2020)

A proof of ∀xA(x) is a construction which transforms every proof that d belongs to the domain into a proof of A(d).

and/or Constructive (Bridges and Palmgren 2018)

A proof of ∀x∈S,P(x) is an algorithm that, applied to any object x and to the data proving that x∈S, proves that P(x) holds.

Universal Generalization.

[103]   Pursuant, by the way, to non-ampliative inference (see e.g. page 71 of Mattessich's *Instrumental Reasoning and Systems Methodology* (Mattessich 2012)) to which such "deductive" systems are necessarily committed.

The bound-variable does not appear in fol with a distinct sense, but as a semantically "synonymous" notational device, syntactically *necessary* (in some instances) "to restrict the scope of generality to part of the formula"; Kleene offers an excellent example (§32, page 151 of reference (Kleene 1971)); while the notations, A(x) ⊢^{x} B & ∀xA(x) ⊢ B, are interchangeable, he shows that "No formula without a quantifier can be written which is synonymous with the formula of step 3, Example 3 (of the conclusion of Example 6.)…, which is confirmed by Frege in terms of (meta) Roman to (object) Gothic letters, on pages 65&66 of his Basic Laws of Arithmetic (Frege 1967): "We write such a *translation* [our emphasis] thus:

"⊢Φ(X)" ⊃ ⊢∀xΦ(x)

…Accordingly, with a Roman letter we cannot ever express the negation of a generality…[so that] the expression of generality by Gothic letters and the concavity does not thereby become superfluous."

corroborating Kleene's finding. See similar line in Frege's Begriffsschrift (Frege 1879), page 25, where the letters are now German and Latin: "This explains why the concavity…"

On Wittgenstein's account, it is "nonsense" to say "we cannot enumerate all the numbers of a set, but we can give a description", for "the one is not a substitute for the other" (Waismann 1979), and an Induction ∀xφ(x) simply *stands proxy* for "φ(0)∩[φ(n)→ φ(n+1)], n-arbitrary" (Wittgenstein 1964, §164), and not for an infinite conjunction, as "there are no such things" (Wittgenstein 1974). As regards Shapiro's dissent (subsection *Ohio, 1998*, paragraph "Shapiro continues,.." above) it seems evitable that someone would come along and tie the mystery of Truth to the mystery of the Infinite; in the study of quantum mechanics, the phenomena is known to as a Minimization of Mystery (Chalmers 1995).

excellent article (Yaqub 2008): one truth definitional (or translational), correct by a borrowed sense (by a thought expressed[104]), the second Platonic, and this third one material, "truth-conditional" (Raatikainen 2008) (Yaqub 2008), correct by the state of affairs referenced, by "the actual way things are" (Glanzberg, Truth/1.1.1 The origins of the correspondence theory 2018), which truth may *only then*, in the usage of the day, be "asserted."[105] Anyway, while the deflationary truth of "2+2=4" rests with its sense, its ontological truth rests with the material conditions sufficient to its satisfaction—the precise material between realists and anti-realists remaining in dispute. Yaqub writes:

One might say that the truth of the statement 'there are three prime numbers between 15 and 25' commits us to the existence of prime numbers. But this is so only if one accepts classical semantics. Someone who is a constructivist might say that the truth of the statement above commits us only to the existence of certain mental constructs. It shouldn't be expected that truth theory would settle the debate between realists and constructivists. A conceptual deflationist would say that the condition under which the mathematical statement is true is that there are three prime numbers between 15 and 25.

This is a distinction that our philosophers overlook, and (Piazza and Pulcini 2015) becomes the latest in a long line of wearisome attempts to enlist the deflationist-of-straw into the ranks of a long-standing anti-realist "truth=proof" cause that is simply not his.[106]

Consider a proposition ϕ. Whether it is satisfied, we consult the world; whether true, a dictionary or translation manual. And yes, in respect of P

---

[104]   Expressing the same thought (Frege, Sense and Reference 1948).

[105]   "…and now Tarskian 'referential satisfaction' or 'denotational semantics'…" might be a better go at it (McCarthy and Fine, Truth without Satisfaction 1984) (Glanzberg, Truth/2.4 Reference and Satisfaction 2018).

[106]   For supplementary reference we cite Glanzberg's Stanford-Philosophy article (Glanzberg, Truth/1.1.1 The origins of the correspondence theory 2018) (sections 4.2, 5.1, 5.2, & 6.3 esp.), and the introduction to McCarthy's "Truth without Satisfaction" (McCarthy and Fine 1984). The inconsistency Yaqub draws between his Conceptual-Deflationism+Bivalence and Metaphysical Deflationism (our anti-realism)—corresponding e.g. to the anti-Gödel-Incompleteness of Vidal-Rosset's Bivalent Internalism (Vidal-Rosset 2006)—though interesting, does not concern the present deflationary question.

$$\text{if } \vDash \phi \text{, then we } \textit{say } T(\phi)$$

while in respect of the world, we say ϕ. It goes without saying then that in explicating the notion of "truth" in the context of denotational semantics, as does Tarski in his "Semantic Conception of Truth" (Tarski 1944),[107] one necessarily employs the language of both meta-P and P—using one, mentioning the other—and so, a rule of translational Meaning, Convention-T, that does not expand on the original ordinary-language sense. Even so, there remain many views on what constitutes truth—how to best speak of it coherently, sensibly; views held for all sorts of reasons. Mathematics-philosopher Peter Smith warns against a tendency current in his trade to doubt the *genuine truth of mathematical facts* (such as "1+2=3") *in the usual humdrum sense* (§5.3 of (Smith 2013)), which one would guess to be the doubt that anti-realist internal facts—mathematical "fictions"—are equally significant, *genuine*, as realist external ones— a question of reference couched in translational terms.[108]

Philosopher Rota notes the same (Rota 1991) and suggests we judge philosophic belief and conviction not by what adherents say, or by what others think they ought to say, but by what adherents do.

Very well. Let us ask, who then doubting the materiality of Gödel's first *bi-domain* Capture on page [182] of the original paper

$$\textbf{vtype(n,x)} \Leftrightarrow \textbf{x is a VARIABLE OF TYPE n}$$

would trudge though seven additional pages to the Undecidability punchline? [109] Who really?

Ninety-years plus, and the row over the meaning of the Gödel sentence muddles on. We may not expect that it will anytime soon end.

---

[107]   Less a theory or definition than a "characterization" of truth (Field 1972).

[108]   The same if more seasoned Smith earlier in section "Indirect Self-Reference" debating a Mr. Jones on an obscure 2006 sci.logic forum page (Jones 2007), today a leading GI authority.

[109]   i.e., otherwise, accepting as equally contentual its P-numerical and metaP-synactical propositions.

IV. Wittgenstein's Perspective

The charismatic Ludwig Wittgenstein wrote over 85 manuscripts in his lifetime, publishing two modest articles and only one book, *Tractatus Logico Philosophicus,* the most condense 90-page text that you will ever read. Yet last year alone, 70 years after his death, Wikipedia records over 600,000 "Ludwig Wittgenstein" page views, and for the same period, according to Google, the name "Wittgenstein" heads titles of more than 500 scholarly articles, so as to seem there might hardly be more about him and his thinking that could be said.

This is no less so of his surprisingly sparse, terse, and often enigmatic comments on GI. Having "solved all philosophical problems" in his *Tractatus*, Wittgenstein maintained an enduring if arm's length interest in the leading mathematics foundational discussions of the day, and in particular in the prominent Program of Hilbert inaugurated that same decade, whose goal had been to establish the consistency and completeness of mathematics so as once and for all to set it on an unshakably firm foundation. And while by this alone Wittgenstein might by 1931 have had a natural interest in GI as well, as indeed was the case, perhaps there too there is hardly more that could be said.

Wittgenstein would return to the world of Cambridge philosophy in 1929, having left there 10 years earlier following publication of *Tractatus*. By then he had given away a considerable inheritance and spent the intervening years in the Vienna countryside working odd jobs. Thus began his mathematical Middle-Period, as it is called, marked by a new finitistic outlook under the influence of Brouwer's Intuitionism and in close acquaintance with Logicist mediator Ramsey, among others. What had not changed in the decade away from Cambridge, however, was his guiding principle regarding the proper business of philosophy, that "philosophy is not a theory, or a doctrine, but is rather an activity. An activity for the clarification of thoughts" (Wittgenstein, Tractatus Logico-Philosophicus 1922, 10).

By contrast, truth values of bonafide propositions were to be left for the sciences to decide. Rather than answering philosophical questions, this principle or way of philosophizing as adhered to by Wittgenstein more readily leads to critique of the question itself—a question more often than not found on analysis to be lacking in clarity, and so in need of re-statement appropriate to some desired linguistic sense, typically by means of definitional, semantic re-assignment. By the "definitional" we mean interdependent-definitions, e.g., those of "truth" and "provability", "propositionhood" and "theoremhood", etc. The idea is that solutions to a genuine philosophical problem may lie in clarification via reworking of the problem itself, that most philosophical problems are mere misunderstandings that "arise when language goes on holiday" (Wittgenstein 1953, §38). And this is certainly so of Wittgenstein's consideration of mathematical foundational problems, such as the question of the completeness and consistency of formal arithmetic. "The problems are solved, not by giving new information, but by arranging what we have always known" (Wittgenstein 1953).

As Wittgenstein saw it, the great majority of what were considered philosophical problems were just of this type[110] and so should be addressed with the special care that does not rush for an answer, but seeks first to dissolve the question away by means of its clarification. Thus 10 years earlier, in one terrifyingly terse Tractatusian verse (Wittgenstein 1922, §3.333), Russell's paradox is simply made to disappear.[111] Such was the analytic manner and style of definitional solution to philosophical questions that set Wittgenstein apart from all others of the day, and it is natural that this would be his initial attitude and approach 12 years later when faced with what he liked to call the Gödelian Paradox.

There were, presumably, elements in Wittgenstein's philosophical purview and range of erstwhile professional associations with at least Logicism, Formalism and Positivism, for example, that on their face might have set him at odds with the conceptual structure of Gödel's result: Logicism, its mathematics being proper subset to logic, and the theorems that constitute its logical calculus, accordingly, being its non-contingent pseudo-propositions,[112] neither genuinely true nor false, except one might say "internally", with respect to the calculus. Truth predications then become superfluous, and when made may only mean that the thing so predicated is "calculable", "deducible", "provable", what have you.

---

[110]   Deriving, typically, from misunderstanding, bad grammar.
[111]   Wittgenstein famously "dispenses with Russell's Paradox" without having to rely on Russell's theory of Types.
[112]   Though Wittgenstein's usage on the subject is by no means consistent.

There is also the real if seldom mentioned all-too-human prospect that Wittgenstein might have taken offense with the very title of Gödel's work, which seems to take a pot-shot at the School and its founder (Wittgenstein's former mentor) already well in decline. Then there is verificationist positivism, for which *meaning* is verification, i.e., proof, internal or not, by means of which Gödel's Incompleteness, $\{T(g) \cap \neg \vdash g\}$, becomes on its face "contradictory", "nonsensical", and so forth.[113]

The only point being that Wittgenstein might well have objected to GI from any number of natural standpoints. Instead, we have a Wittgenstein who misreads the Gödel sentence as a linguistic *false-cognate* in two distinct languages by which these meanings conflict; what Hofstadter above likens to a double-entendre:[114]

> I imagine someone asking my advice; he says: "I have constructed a proposition (I will use 'P' to designate it) in Russell's symbolism, and by means of certain definitions and transformations it can be so interpreted that it says 'P is not provable in Russell's system'. Must I not say that this proposition on the one hand is true, and on the other hand is unprovable? For suppose it were false; then it is true that it is provable. And that surely cannot be! And if it is proved, then it is proved that is not provable. Thus it can only be true, but unprovable."
> Just as we ask, "'Provable' in what system?", so we must also ask, "'true' in what system?" 'True in Russell's system' means, as was said: proved in Russell's system; and 'false in Russell's system' means: the opposite has been proved in Russell's system.—Now what does your "suppose it is false" mean? In the Russell sense it means 'suppose the opposite is proved in Russell's system'; if that is your assumption you will now presumably give up the interpretation that it is unprovable. And by 'this interpretation' I understand the translation into this English sentence.—If you assume that the proposition is provable in Russell's system, that means it is true in the Russell sense, and the interpretation "P is not provable" again has to be given up.[...]

---

[113]   And, mathematical-proposition ≡ theorem, with "meaning" as method of *positive* verification. Or "Meaning" *particular to* the calculus in which the verification is made (Wittgenstein 1964) .... Meaning *demonstrated by* actual proof, as in (Wittgenstein 1974, §32):
"the proposition $f(x)$ holds for all cardinal numbers" means "it holds for $x=1$, and $f(c+1)$ follows from $f(c)$".
so that the meaning of a generalization is for Witt synonymous to, indeed *defined as*, its inductive proof.

[114]   See above sub-section *What does* g *not say.*

This is Wittgenstein's "Notorious Paragraph", which at the time of the publication of his *Remarks on the Foundations of Mathematics* (Wittgenstein 1978), in which it appears, drew such pointed and unanimous professional criticism as to permanently damage his good name.

Victor Rodych (Rodych 1999), for one, takes it that the critics have overreacted, and urges that we now ignore or at least not weigh too heavily Wittgenstein's careless though really minor misstep—a misstep for which Gödel almost certainly shares blame: Gödel, for his part, outlining in the Introduction of his paper a semantic derivation never followed, and for purely rhetorical effect comparing GI with Richard's paradox; Wittgenstein, for his part, probably not reading beyond that Introduction. Rodych urges that we not judge on the "expert testimony" of authority alone but reconsider Wittgenstein's *Remarks* rather in the light and context of his broader conceptual framework.

As to the merits of such an approach, opinions vary, but the matter of Wittgenstein's approach to Gödel's work has been and remains of such high general and volatile philosophical interest that it is not uncommon to find a mathematician here taking one side of the issue and there taking the other.[115] The surprisingly small corpus in which we find Wittgenstein's words on Gödel—indeed, only a few passages embedded in large works—are pored over, reviewed again and again, interpreted and re-interpreted by successive generations of aspiring students of mathematical philosophy, and there they maintain, at least up to now, a dedicated readership all their own.

Toronto, 1988

York Psychology

Among those of note who would heed Rodych's call are Steiner (Steiner 2001), Priest (Priest 2004), and Berto (Berto 2009), though none perhaps more strident than his York psychologist colleague Stuart Shanker, who's *Wittgenstein's Remarks on the Significance of Gödel's Theorems* to *Gödel's Theorem in Focus*, edited a decade earlier by himself (Shanker 1988), had probably been its inspiration. On the opening pages he writes:

---

[115]   Wittgenstein's former student Kreisel, for example.

> The primary aim [here] has been to locate Wittgenstein's remarks on Gödel's theorem within the larger context of his approach to the consistency and decision problems in order to show both how these obscure fragments fit into the overall strategy of his proposed resolution of the foundational crisis.

The "larger context" derives out of similarities and differences between Logicist and Positivist conceptions of the proposition, logical inference, and mathematical equations. Much as in Russell & Frege's Logicism, Wittgenstein conceives

The Evening Star is the Evening Star

as a statement belonging to logic, its correctness, *validity*, being a matter of form, syntax, and as such is non-contingent on the world of matters of fact. Such a proposition is said by consequence to lack *genuine* sense. On the other hand a statement

The Evening Star is the Morning Star

is decided true or false by correspondence with the world of facts, and thereby contingent upon its relevant states of affairs.[116] The first is said to be a *logical inference*, and only the second a genuine *proposition*.

There are then the statements of Mathematics which at least for Wittgenstein occupy a category all their own. Their correctness, if not truth, like that of the logical inference, being identical to their purely syntactic *provability*. Unlike the logical inference, however, the mathematical proposition is endowed with meaning, with a sense, as its *proof* or *refutation*. Having neither the essential property of the proposition nor quite that of the logical inference, the mathematical equation in Wittgenstein's usage is said to be a *pseudo*-proposition, numbers being unique (*sui generis*) mathematical objects.

And it is such a conception of the mathematical statement that it might put Gödel's results under suspicion on at least two counts: the theorem itself[117] seems to reduce or *collapse* to contradiction or paradox

$$GI \Rightarrow [\,\{T(g) \cap \neg \vdash g\,\} \cap \{T(g) \Leftrightarrow \vdash g\}\,] \Rightarrow \{T(g) \cap \neg T(g)\} \equiv \perp$$

---

[116]   Or, while the physicist does not consult the world as to whether the work done by a force equals its path-integral, he does as to whether it also equals the kinetic energy it imparts to the object it acts on along this path.

[117]   That there is a mathematical proposition both true and not provable; i.e., both true and *false*.

while its driving pseudo-propositional Gödel-sentence (having no proof being tantamount to having no meaning; being "a meaningless concatenation of symbols") is reduced, or *collapses*, to meaninglessness, nonsense.

$$GI \Rightarrow \neg\{T(g) \cup T(\neg g)\}\ {}^{118}$$

g being neither true nor false.

Then there is the Hilbert Program in whose conceptual framework Gödel does his work. In the capacity of a paradox Gödel's theorem serves then as a reductio ad absurdum of the Program itself (much as Russell's Paradox had demanded a Fregean Logistic rethink) and as good a reason as any to abandon the project altogether, for which Wittgenstein *is said* to have long advocated.

And here is where Stuart Shanker steps in, employing an impressive range of rhetorical devices in defense of this framework. In the middle of a randomly selected page 178 of reference (Shanker 1988) we find

> ...Wittgenstein's repudiation of the need for a consistency proof turned on the argument that Hilbert's Programme arises from a skeptical concern which is *ab initio* unintelligible; it introduces a spurious epistemological dimension into an issue which is solely concerned with the logical grammar of mathematical propositions. And since doubt can only exist where a question exists; a question can only exist where an answer exists, and this can only exist where something *can* be *said*' (NB 44),[49] the proper response to Hilbert's problem is to *dissolve* it by demonstrating that Hilbert's worry, far from being irrefutable, is logically excluded by the normative character of mathematical propositions.

---

[118]   Thus, by/in Wittgenstein's conceptual framework, Gödel's Incompleteness Theorem, Consistency and PR-completeness collapses to an Incompleteness Paradox and Gödel sentence meaninglessness:

Witt: $T(g) \Leftrightarrow\ \vdash g$

Diag: $T(g) \Leftrightarrow \neg\vdash g$

Con(P): $\neg[\vdash g \cap\ \vdash \neg g]$

collapse to the non-propositionhood of g:

$T(g) \Rightarrow_w\ \vdash g$

$T(g) \Rightarrow_D \neg\vdash g$

$T(g) \Rightarrow \perp$

$\therefore \neg T(g)$

$T(\neg g) \Rightarrow_w\ \vdash \neg g \Rightarrow_{c} \neg\vdash g \Rightarrow_D T(g) \Rightarrow_w \vdash g$

$T(\neg g) \Rightarrow \neg\vdash g \cap\ \vdash g \Rightarrow \perp$

$\therefore \neg T(\neg g)$

$[\neg T(g) \cap \neg T(\neg g)] \Rightarrow \neg\ [T(g) \cup T(\neg g)]$

meaning no more than that Hilbert's unanswered consistency question, his *skeptical concern* whether the Peano axioms are self-consistent, cannot intelligibly be stated[119]—a *concern* that within a page becomes a "skeptical crisis", and on the next, a "skeptical dilemma",[120] which, because Gödel, with the rest of the world of practicing mathematicians, imagined that he had *understood* this "skeptical illusion" (ibid, 181), thought it "cogent",[121] a page later it becomes a "skeptical dilemma" posed by GI itself:

> [We see that Gödel's Incompleteness] is not a refutation, but rather as a *reduction ad absurdum* of Hilbert's programme. If Gödel's theorem does land us in a skeptical dilemma[59] then this can only mean that the premises which initiated this line of thought must themselves be *au fond* unintelligible (ibid, 182).

And the whole of Shanker's article runs much like this—hyperbole, paradox, and loose connotative associations that on every other page take favorite aim, variously, at the "sense", "meaningfulness", "intelligibility", "coherence", "comprehensibility",... what have you, of the Gödel sentence *g*. We offer a representative taste of what the reader can expect.

> Gödel's proof serves as a *reductio ad absurdum* of the premise that philosophical problems can be solved mathematically (ibid, 184).
> ... A mathematical conjecture is meaningless until the logical connections articulated in its proof has been forged (ibid, 185).
>
> ... to be a *meaningful* mathematical problem is ipso facto to be solvable, and it is for that logico-grammatical reason that 'There is no *ignorabimus* in mathematics.' (ibid, 186).
>
> ... [Gödel's Incompleteness] literally demonstrates that if one takes the unproved expression and treats it as if it were a theorem, a contradiction results (ibid, 189).
>
> ... In any given case the elucidation of a theorem must proceed from and be strictly confined to the domains of the system in question (ibid, 190).

---

[119] A notion whose own intelligibility 70-years-on remains "controversial and illusive" (Schönbaumsfeld 2016).

[120] A well-known philosophic metonym (Contributors, Metoymy 2021) whether or not to accept a strict Pyrrhonian Principle (Contributors, Pyrrhon of Elis 2021), which suspends judgement with regard to uncertain knowledge i.e., a genuine dilemma whether or not to accept imperfectly knowable knowledge and not a by-word or trope for any and every unanswered yes/no question.

[121] And indeed, arithmetically material i.e., the question whether or not prov(0=1).

... The proof of provability differs from the proof of a proposition (ibid, 191).

... There cannot be a hierarchy of proofs.... There can't be such a thing as meta-mathematics. (ibid, 192)

... Mathematics cannot be incomplete; any more than a sense can be incomplete (ibid, 198).

... Hilbert was only led into his illicit categorical distinctions because he persisted in searching for ways to preserve the referential conception of mathematical propositions. Remove this framework and the need for a consistency proof does not occur (ibid, 210).

... If I'm clear about the nature of mathematics, the question of consistency can't arise at all (ibid, 215).

... The proof [of GI] only proves what it proves.... The Gödel numbers are not numerals which denote expressions in an "object calculus" (ibid, 216).

... [We] cannot draw any philosophical consequences from Gödel's formal (two-tier) version of the pseudomenos because it is incoherent to assert that "P is unprovable and true" (ibid, 221).

... [The] meaning of a mathematical proposition is strictly determined by the rules governing its use in a specific system (ibid, 229).

... [To] describe a mathematical expression as unprovable is to deny that it is a mathematical proposition; i.e., that it is unintelligible (ibid, 230).

... [The] philosophical substance of Gödel's argument is contained in the preface and not in the body of the proof.... Gödel's 'informal introductory arguments' have quite pronouncedly displaced the formal proof as the focus of philosophical investigation (ibid, 232).

... Gödel's theorem serves as a reduction ad absurdum of the philosophical framework underpinning Hilbert's programme: viz, by forcing us to accept that a mathematical theorem could per impossible be intelligible prior to or independently of the set of rules which create its meaning.... [If GI] is purged of Hilbert's epistemological bias a formidable mathematical analogue of Wittgenstein's philosophical critique of Hilbert's programme emerges (ibid, 234).

... Wittgenstein set out to remove the *Bedeutungskorper* conception of mathematical meaning, for it is this picture that sustains Gödel's interpretation (ibid, 237).

The suggestion of some tantalizingly refined if impending sense that runs throughout these figurative passages—like the first, and perhaps like Wittgenstein's own thinking on these matters—rarely take full form before Shanker has moved on to the next metaphor. Whether the views they express are those that had been held by Wittgenstein prior to his knowledge of Gödel's work, as claimed,[122] or are of a post-GI vintage— i.e., whether views held on principle vs. ad-hoc—we leave for the reader to judge.[123]

What in any case is striking throughout Shanker's representations and necessary for his Wittgenstein arguments to begin and proceed at all, is that the truth/provability analytic equivalence therein insisted upon in one context is taken to be epistemically synthetic in another.[124] Taking another look at our above summary of Wittgenstein's framework in which GI and its Gödel sentence are presumed to collapse, respectively, to the paradox and nonsense referred to in those passages, we see that a synthetic relation between statements true and those provable given there

$$T(g) \Leftrightarrow \neg \vdash g$$

does not properly render the Internalist conception of mathematics. No. The point pressed by Internalists, rather, is that the mathematician is willing say that some φ is *true* only when he has proven it (that being what the predication "is true" to him *means*[125]), and that it is best that the term be altogether eliminated from the discussion of mathematics. For the In-

---

[122]   Which proposition Shanker fully recognizes as crucial to the credibility of Wittgenstein's avowed motives: "The origin of Wittgenstein's approach to Gödel's theorem lies thus before the announcement of Gödel's result: a point of utmost importance that has hitherto been ignored" (ibid, p 192).

[123]   One might look into the *specific cited support* Shanker offers for his claim on page 192 that "The origin of Wittgenstein's approach to Gödel's theorem thus lies before the announcement of Gödel's results" and on page 236 that "..the framework of...Wittgenstein's remarks... was already in place before the announcement of Gödel's theorem." On the other hand, consider Viktor Blasjo's review (Blasjo 2008) which characterizes Shanker's own "Gödel's Theorem in Focus" article as an unsuccessful attempt to piece together Wittgenstein's "...incoherent scraps ....", while "abusing his editor's privilege", etc. In precisely what, one may ask, do these scraps consist? Mainly brief anaphoric excerpts from Wittgenstein's 1929-30 *Philosophical Remarks* (Wittgenstein 1964), (with a few from 1931-33 *Philosophical Grammar* (Wittgenstein 1974)). Though Shanker cites "sections" §108, §109, §114, §121, §149, §152, §153, §154, §157, §158, §163, their thinly relevant passages are perhaps enough to fill a paragraph. At least one Shanker reviewer finds herself persuaded by the sheer "plethora of citations from those works" (Matthíasson 2013).

[124]   Synthetic, as among conceptually independent ideas.

[125]   An ordinary-language colloquialism then, used when speaking to a general public or beginning students.

ternalist, truth is not material but redundant to provability, which in Kant-
ian terms could be put that the relation is analytic, of the nature of a definition, or in today's Communicationist terms, that it is not theoretical, as
"between distinct concepts", but is rather an "existence statement" that
makes a "definition" (Reynolds 1971).[126]

Let us ask then, why do we say that the Gödel sentence is true? And
where is this supposed to be shown? Diagonalization of the Gödel sentence takes place after all in meta-P,[127] or equivalently, if formally, in the
weakest, lowest P-progression in Turing's transfinite recursive hierarchy,[128] and so is rendered by the Wittgenstein Internalist

$$\neg \vdash_P g \Leftrightarrow \vdash_{\text{meta-P}} g \qquad\qquad \text{meta-meta-P}$$

or equivalently, in terms of Turing's extended $\tilde{P}$ [129]

$$\neg \vdash_P g \Leftrightarrow \vdash_{\tilde{P}} g \qquad\qquad \text{meta-P}$$

in which GI itself becomes

$$\text{Con(P)} \Rightarrow \{ \neg \vdash_P g \cap \vdash_{\tilde{P}} g \}.$$

In this framework, there is no Hilbertian completeness problem, Wittgenstein having deprived us of the means to even state it, for which we perhaps ought to be grateful. Yet the system we are left holding feels no less
incomplete. For indeed the substance of the question does not go away;
it has a different face, a form as yet unnamed. Our original system P remains "incomplete", in the relative sense that its extension, $\tilde{P}$, asserting
also our commonly-held conventional confidence in the consistency of
the P-calculus, is *more* complete—an axiomatic confidence, one might argue, implicit to our very acceptance of the Peano axiom *set* in the first

---

[126]   So that the predication itself is defined out of the language. Rodych himself later ad-
mits (Rodych 2018):
This, in a nutshell, is Wittgenstein's conception of "mathematical truth": a true proposition of *PM* is an axiom or a proved proposition, which means that "true in *PM*" is
identical with, and therefore can be supplanted by, "proved in *PM*".
and (Rodych 1999):
We speak of 'true' and 'false' mathematical propositions, but we do not need 'truth'
and 'falsity' in mathematics—they are completely eliminable.

[127]   See section "Königsberg, 1930", illustration (B).

[128]   See definition "(ii)" above in section "Princeton, 1962".

[129]   Recall, $\tilde{P} \equiv \{P \cap \text{Con(P)}\}$

place, so that the method of mathematical formalization starts out, one might say, irrevocably steeped in Truth.

This forms part of Gödel's own argument for the Inexhaustibility of Mathematics; for as the man who "perceives the axioms under consideration to be correct... also perceives (with the same certainty) that they are consistent" (Gödel and Feferman 1986, 309). It is the same quite natural sense in which one would say that Peano Arithmetic is more complete than Robinson's, while at the same time, the expression "x+y = y+x", unprovable in Robinson's, is regarded by no one a "meaningless concatenation of symbols." Indeed, we see that Wittgenstein's conception held faithfully to —refraining e.g. from using terms outside its own language-game— far from disproving it, provides an alternative framework by which to state Gödel's Incompleteness.[130]

But "such a proposition is 'true' or 'false' in a different sense from a proposition in, or better, of PM" Rodych warns.[131] In one breath we are told then that the Wittgensteinian Verificationist meaning of a mathematical proposition is its proof, while in another that its Wittgensteinian Internalist meaning is distinct to the proofing system. So that on the one hand "2+2=4" may have multiple meanings in P, and on the other, these

---

[130]   On the other hand, when we commix and so confound these terms so as to produce a locution "true but unprovable sentence"—i.e., confound them in the manner we have seen—the false-discord that brings us to the present discussion is readily manufactured, as witnessed here in Berto's brief pro-Wittgenstein "exposition of the First Theorem" (Berto 2009, p 11):
Since g "claims" (via arithmetization) to be not provable, and we have just proved that it is not provable, then g just is what it claims to be; hence, it is true. However, this simple reasoning cannot be performed within the theory: the truth predicate for PA, were it expressible within PA, under the usual conditions would originate the Liar paradox; whereas the provability predicate is expressible. Gödel himself pointed at the analogies between his undecidable sentence and such paradoxes as Richard's, or the Liar. However, it seems clear that, whereas the Liar sentence, "This sentence is false", produces an antinomy (if it is true, it is false, and if it is false, it is true, therefore—given Bivalence—it is both true and false), with the Gödel sentence, metamathematically read as "This sentence is not provable", no contradiction is forthcoming.
With "the Gödelian results" then—having against Wittgenstein's Internalist objections established "a fundamental gap between provability and truth"—as "one of the great moving forces behind the modern resurgence of Platonism".

[131]   "For example, the most obvious way to make the Gödelian proposition 'P' proved (or provable) is to simply add it to the axiom set of Peano Arithmetic (PA), such that we create a new calculus, call it "PA + P". Now, ' P ' is not provable in PA, but it is provable in PA + P. As Wittgenstein says in #7, such a proposition 'is "true" or "false" in a different sense from a proposition in, or better, of Principia Mathematica." (Rodych 1999, p 181).

differ from those in P̃, even though their deductions, Gödelian "proof-figures", are identical... It is a line that reviewers repeat (by all signs, ob-liviously) over and over again (Matthíasson 2013):

> If a sentence is internally related to its proof (i.e., as Wittgenstein maintained, the proof establishes the very meaning of the proved sentence), then it is not possible for the same sentence (that is, for a sentence with the same meaning) to be unde-cidable in a formal system, but decided in a different system (the meta-system).

and:

> According to him [Wittgenstein], the meaning of a mathematical sentence is deter-mined by the rules that govern its use in the calculus, and in particular by its own demonstration (which is why a demonstrative incompleteness in the theory would become *eo ipso* an incompleteness of meaning.

To be clear, Wittgenstein remark in the #7 *Remarks* passage cited by Rodych, "But may there not be true propositions which are written in this symbolism, but are not provable in Russell's system?... Why should not propositions—of physics, e.g.—be written in Russell's symbolism?" (Rodych 1999, 180) refers to and does indeed rightly hold for systems of a common script whose languages are inter-translatable, such as Gödel's system P and its G-transformation P' (or for that matter, systems such as the Peano axioms implemented in Fortran and in C#), and not for such systems P and its *extension* P̃ (or for Fortran implementations of Peano vs. Robinson Arithmetic, as in our earlier example) as Rodych's suggests, with all the misdirection of a mixed-metaphor, the fallacy of a *false-com-parison* (Contributors, False Equivalence 2021).

Were we on the other hand to reason in the vein of Wittgenstein's perhaps more accessible ordinary-language analogy, we might agree that yes, given the common script, and the fact that it constitutes a word in the Latin language, it is elementary that this does not make the concate-nation "mutatus" a word, and so with *meaning*, in the English language—no more than Newton's second Law coded in binary or the P' statement, G[2+2=4], are statements in P. It is also true however that the neologism "laser", with a thousand others, would carry no meaning for the pre-1960s speaker of English, whereas the commonly understood meaning of the word "hat", with most others, has since then not changed much. Pre-1960s English usage, then, we might find it in some ways inadequate for our needs today; one might even think of it as being *incomplete*, as indeed Gödel would call P in respect of $\tilde{P}$.

It is not that g in Wittgenstein's usage (or framework) has in P a meaning that differs from that in $\tilde{P}$—no. For in P there is no proposition g; it does not appear in the calculus (i.e., as a theorem), so that in this framework there is nothing whatever via P to say of the expression g. The most relevant PR §153 reference Rodych quotes, "A mathematical proof is an analysis of a mathematical proposition. It isn't enough to say that p is provable, we have to say: provable according to a particular system. Understanding p means understanding its system."—clearly overshoots the verificationist mark, and ought to be conditioned, "to the extent that the 'particular system' affects the yet particular proof of p."

For it goes without saying that it is superfluous to the verificationist principle that the meaning of some proposition p should depend also on parts of the system not employed in its verification—mathematically verificational sense being proof-specific, and not specific to the proofing system. Steiner (Steiner 2001) observes:

> Wittgenstein writes as though Gödel had constructed a trivially new interpretation of the 'Russell' formalism—like calling a tail a leg, and then saying that a cow has five legs. And therefore, Gödel was free to call anything he wanted 'true' (ibid, 274).
> ... The 'picture' of truth as the winning position of a game suggests to him the following idea: extending the 'Russell system' by adding undecidable sentences is like defining the rules of a new game, and is not like discovering new terrain (ibid, 275).
> ... Wittgenstein's observation
>> Just as we ask: " 'provable' in what system?", so we must also ask: " 'true' in what system?"
> —is very misleading, because what is true in one system may be mathematically fixed by what is provable in another. The appropriate analogy here is to the extension of theorems about the reals to the complex plane, as explained above. Mathematicians had no real choice in the extension; there was essentially only one way to do it, if it were to be done at all (Steiner 2001, 278).

Furthermore, only for those expressions that appear in no mathematical calculus whatever, *and solely for these*, does the pre-Gödelian Wittgenstein reserve the designation "meaningless"—for example, such expressions as those corresponding to the conjectures of Goldbach and Fermat. If however we are confident or conditionally willing to accept the mutual consistency of the Peano axioms (which, again, is already implicit to acceptance of the set) by consequence of which P does not prove "1= 0"— to that *axiomatic* extent—we may make whatever application or *use* of the *mathematical statement* Con(P) ($\equiv \neg$prove(1=0)) as we wish, just as we do of the statement "2+2=4", and from which we may draw what at least the *latter* Wittgenstein calls a "meaning".

GI then poses no more an issue with Internalist nor Verificationist mathematics (nor paradox in or reductio of Hilbert's work) than the fact that, "What is called losing in chess may constitute winning in another game" as Wittgenstein puts it (Wittgenstein 1978, p 51).[132]

Toronto, 1999

York Philosophy

For his part, Rodych in *Wittgenstein's Inversion of Gödel's Theorem* is perfectly happy to corroborate colleague Shanker's enthusiastic if shaky reading of Wittgenstein's *pre-Gödelian Philosophical Remarks*,[133] in support of the thesis that mathematical "propositionhood" for Pre-Gödelian Wittgenstein is synonymous with "theoremhood", their tenuous timeline and the quotations themselves only highlighting, again, the need to establish some credibility and good-faith for Wittgenstein's arguments (i.e., his motives) that may be called into question if only by their hostile tone.

The "Inversion" referred to in Rodych's title and finally unveiled on page 25 of the article turns out to be no more than GI's standard reading *in contraposition*

$$\neg[(\text{Con}(P) \Rightarrow \neg \vdash g) \Rightarrow \neg \vdash g]$$

with Rodych's explanatory assurance that

> Wittgenstein's main aims in his RFM remarks on GIT is to remind us that GIT in no way precludes a derivation of 'P' in accordance with the rules of "Russell's system", since Gödel (only) shows that "If" Russell's system is consistent, then 'P' is not derivable".

which continues

Contrary to the claims made by Paul Bernays and Georg Kreisel (Section 2.3), I shall argue that Wittgenstein does not overlook or underappreciate the 'premise' or "assumption of consistency" in GIT, but rather

---

[132]   See also Yaqub's observations (Yaqub 2008) in section "Lisbon & Pisa, 2014" above.
[133]   Citing from there three times as many passages.

inverts GIT to show that a proof of Gödel's "true but unprovable" proposition within "Russell's system" has in no way been precluded.

Wittgenstein's inversion "argument" then is more of a friendly "reminder" that Gödel has only conditionally established the non-provability of g in P. I.e., it reminds us that

$$\neg[(\dot{C}on(\dot{P}) \Rightarrow \neg\vdash g) \Rightarrow \neg\vdash g]$$

so that

$$\vdash g \Rightarrow \neg\neg\vdash g \Rightarrow \neg Con(P).$$

Thus, Rodych recommends that the main aim of Wittgenstein criticisms against GI is only to *remind us* that

$$\neg [(A \Rightarrow B) \Rightarrow B]$$

only if A has not been established. Whether Rodych manages to maintain an inscrutable poker-face in claiming this is anybody's guess...[134]

Of slightly greater novelty, if not mathematical interest, is the challenge Rodych lays down to the reading of GI that has g both true and unprovable. It rests on Wittgenstein's outright rejection of the propositionhood (or even pseudo-propositionhood) of true negated mathematical statements; i.e., of equation *denials*, among which is Gödel's primitive recursive, $R(x) \equiv \neg prov(x, G[g])$[135] for any x whatever, even *before* its generalization to a material identity with the actual Gödel sentence, $g \Leftrightarrow \forall x.R(x)$. Rodych explains:

> In closing, I want to suggest a challenge to this claim that is related to another avenue of Wittgenstein's thought, which can easily be seen as a (non-chronological) continuation of his view of mathematical 'assertions' and his challenge of GIT.

He argues that the absence of a *picture* corresponding to R(1) corresponds to the absence of a "genuine" sense. For while to "John is wearing

---

[134] Of slightly more interest or more value is Rodych's observation of how Witt thinks that

i) $\vdash g \Leftrightarrow \neg Con(P)$

derives from and cannot be established without

ii) $T(g) \Leftrightarrow \neg\vdash g$

however it is read. While Gödel actually derives i) in AA, *then* ii) in BB, one mutually independent of the other.

[135] introduced in section *Königsberg, 1930*, above

a green hat" there corresponds a definite picture (which, depending upon the facts, may or may not then be negated), to the statement "¬(5×5 = 30)", there does not; a circumstance in which "5×5 = 30" can simply not be imagined, so that "...negation means something different in arithmetic from what it means in the rest of language".

The difficulty does not however prevent Rodych from discovering the meaning of "¬(5×5 = 30)", or rather from redefining it, in classic Wittgensteinian style. For there, the negation may be regarded as only shorthand for the *positive* disjunction:

"(5 x 5) + 1 = 30 v (5 x 5) + 2 = 30 v (5 x 5) + 3 = 30 ... v (5 x 5) + n = 30 v (5 x 5) - 1 = 30 v (5 x 5) - 2 = 30 v (5 x 5) - 3 = 30 ... v (5 x 5) - n = 30" for a "sufficiently large" n - where n is, e.g., a function of the numbers of digits in the respective numbers)

in which (5×5 = 30) is absent. The maneuver is not however available to Gödel, since "R(1)", we are told, "is itself negative, and so cannot correspond to a fact,"—an explanation endorsed nowhere in the (Wittgenstein 1964, §200) cited reference—referring, presumably, to the {1+n} sequence over ¬prov(1, G[g]). In the haste of excitement Rodych has overlooked what by his innovative "eliminative procedure" is actually the relevant sequence: {G[g] +n} over ¬prov(m, G[g]), for P' proof-figures "m".[136]

In any case, in Rodych's subsequent article commissioned by Stanford four years later (Rodych, Wittgenstein's Philosophy of Mathematics 2018), the argument is entirely dropped.

The article ends with an appeal to the reader's higher philosophical angels, urging that we not "simply dismiss Wittgenstein's remarks on Gödel's First Incompleteness Theorem on the 'expert testimony' of early expert reviewers"—an appeal that first appears in the abstract of the paper. In this same closing paragraph Rodych praises that "the real merit of the various strands in Wittgenstein's remarks on GIT is that they force us to question the meaningfulness of sentence constructions such as 'P.'", and

---

[136]   Here, Rodych has surely forgotten that the statement R(1) can be no less meaningful than the statement that "1 is not divisible by G[g]", captured by the equally primitive recursive, ¬(1|G[g]), and that whatever eliminative procedure we find applicable to one may also be applied to the other. It is also worth mentioning that "1" in ¬proves(1, G[g]) is no P' proof-figure...

in fact, he goes on to admit, making this meaningfulness "highly question-able". The reader must ask then, upon what strength is this question "forced"? On the questionable argumentation presented in its defense, which a paragraph earlier we are asked to excuse for being only "tenta-tive"? On the *expert testimony* of its author?

Cambridge, 2011

Oxford press

Rodych's tolerance—the exceedingly deferential tone and attitude by which the substance of his review proceeds—while perhaps disappoint-ing, comes as little surprise given the aforementioned reputation of and reverence for Wittgenstein, and it is hardly worth mentioning that had the great philosopher's wild (Kreisel 1958, 153), sometimes flippant ar-guments (Dawson 1997), as those above, been made by anyone else, they would in all likelihood have been ignored or dismissed out of hand. "These are hard words to use of a major philosopher", Anderson (below) warns in the introduction of his own early 1958 review of Wittgenstein's GI comments (A. Anderson 1958), itself reviewed in our next section.[137] Perhaps. But the question then becomes, how far may we extend the sug-gested benefit of doubt without overly compromising critical review of his own words? To the extent that Rodych does? Shanker?

Michael Potter's 2011 Oxford article *Wittgenstein on Mathematics* (Potter, Wittgenstein on Mathematics 2011) offers a sober-headed coun-terbalance to what we have so far seen here in this regard. The review covers a broad range of topics of interest from Wittgenstein's middle pe-riod, 1929-36/7, laying out his well-known argument against infinite to-talities, his argument that there is no problem of inconsistency until we actually find an inconsistency, as well as the verificationist argument most relevant to that against the conventional reading of GI, which has the meaning of a mathematical statement being its proof—all, nothing new to be found in such a review. What we *do* find in it new or at least very rare is a frankness, a subtle though persistent tone throughout that

---

[137]   "If true," he goes on, "they should be substantiated as fully as possible…", which on one hand seems appropriate enough, given no more than the standing of the man and the recognized specialty of his subject.

demands clarity of its subject and shows little patience for vague ambiguously enigmatic assertions that promise deep hidden meanings. We find it new and refreshing.

Wittgenstein's answer to the concern that mathematical propositions in his system are subject to multiple meanings, Potter finds "enigmatic". He finds "unclear" Wittgenstein's explanation why undiscovered contradictions of formalized arithmetic can do no harm, "very strange indeed" Wittgenstein's assertion that attributes "the existence of hidden contradictions to ambiguity in the rules", and he doubts whether by 1939 Wittgenstein yet maintained "the doctrine that the meaning of an arithmetical generalization is its proof". On these as on so many points, Potter confirms what technical knowledge and common sense long denied already tells us.

The contrast of his treatment with those of both Rodych and Shanker is striking. Consider Wittgenstein's approach to unanswered mathematical questions or unsolved mathematical problems, such as whether the Goldbach's Conjecture, "every integer greater than 2 can be written as the sum of three primes" generally holds. Shanker and Rodych quote:

> ...Wittgenstein asserts that uncontroversial mathematical conjectures, such as Goldbach's Conjecture (hereafter 'GC') and the erstwhile conjecture "Fermat's Last Theorem" (hereafter 'FLT'), have no sense (or, perhaps, no determinate sense) and that the unsystematic proof of such a conjecture gives it a sense that it didn't previously have (PG 374) because
>> it's unintelligible that I should admit, when I've got the proof, that it's a proof of precisely this proposition, or of the induction meant by this proposition. (PR §155)
>> Thus Fermat's [Last Theorem] makes no sense until I can search for a solution to the equation in cardinal numbers. And 'search' must always mean: search systematically. Meandering about in infinite space on the look-out for a gold ring is no kind of search. (PR §150)
>> I say: the so-called 'Fermat's Last Theorem' isn't a proposition. (Not even in the sense of a proposition of arithmetic.) Rather, it corresponds to an induction. (PR §189)

This is cited as a plausible enough account of the middle Wittgenstein's usage when discussing the matter. But the reluctance or refusal of either Shanker or Rodych to offer even the most elementary critical commentary on so questionable a position—their readiness, eagerness, to grant an open-ended benefit of doubt—clarifies nothing and amounts to tacit ascent to, if not endorsement of, Wittgenstein's unconventional, misleading usage.

And further down, Rodych explains, paraphrasing Wittgenstein:

> The point here is *not* that we need truth and falsity in mathematics—we don't—but rather that every mathematical proposition (including ones for which an applicable decision procedure is known) is *known to be part of a mathematical calculus*.[138]

Compare this now with Potter's account:

> The key distinction for Wittgenstein at this point seems to be whether or not there is a decision procedure for the problem (cf. Säätelä's chapter in this volume). If not, then according to him we do not really understand the problem (despite appearances to the contrary).
>
> > This boils down to saying: If I hear a proposition of, say, number theory, but don't know how to prove it, then I don't understand the proposition either. This sounds extremely paradoxical. It means, that is to say, that I don't understand the proposition that there are infinitely many primes, unless I know its so-called proof: when I learn the proof, I learn something completely new, and not just the way leading to a goal with which I'm already familiar. But in that case it's unintelligible that I should admit, when I've got the proof, that it's a proof of precisely this proposition, or of the induction meant by this proposition. (PR, 183)
>
> Only where there's a method of solution is there a problem (of course that doesn't mean 'Only when the solution has been found is there a problem').
>
> That is, where we can only expect the solution from some sort of revelation, there isn't even a problem. A revelation doesn't correspond to any question. It would be like wanting to ask about experiences belonging to a sense organ we don't yet possess. Our being given a new sense, I would call revelation. (PR, 172)
>
> Every legitimate mathematical proposition must put a ladder up against the problem it poses, in the way that 12×13=137 does—which I can then climb if I choose. (p. 129) This holds for propositions of any degree of generality. (N.B. there is no ladder with 'infinitely many' rungs.) (PR, 179)
>
> Wittgenstein's view leads, then, to the conclusion that we do not really understand Goldbach's conjecture. He has nothing very convincing to say about what mathematicians are doing when they try to prove Goldbach's conjecture.

---

[138] Not bothering in the margins to even mention that the English language is already equipped with a word, a *name*, for that deductively grammatical category; i.e., 'theorems'.

Whatever the final judgement, whatever we may all find lacking in them, Wittgenstein's promoters will have to make a stronger, clearer case to the ordinary reader of English mathematics for the merits of a usage in which Goldbach's conjecture, Fermat's, and the proposition

There are no more than three consecutive 7's in the decimal expansion of π.

all lack grammatical sense.

New Haven, 1958

Language Games

When we admit negation into mathematics as by the expression, ¬(2+2=5), we should not be surprised to discover meaningless "concatenations of symbols" such as the Gödel sentence. For "that", Rodych adds, "is what comes of allowing such constructions." The quotation is from Wittgenstein's reference (Wittgenstein 1978, §11), and in the hands of Rodych concludes no fewer than four distinct lines of argument against GI. Each might simply be countered: No, *that*, rather, is what comes of confounding the denotations of one conceptual framework, one ontology,[139] one language game (Wittgenstein, Philosophical Investigations 1953) with those of another; that is what comes of indulging in or excusing the abuse of words (Locke 1975, 653). How would *anyone* understand it as anything else? How could the preeminent philosopher of language of our 20th century have? The question is rarely asked and is certainly not the sort of staple in the ongoing concern over Wittgenstein's views on Incompleteness that it perhaps should be; one would almost think that no-one had even noticed the quandary or puzzlement of which it speaks.

What most exponents and critics alike *have* however noticed and generally agree upon is that given the novelty of Gödel's mathematical approach and the timing of Wittgenstein's return to mathematical philosophy, Wittgenstein had simply, perhaps carelessly, misread the newcomer's line of deduction, mistaking the Gödel sentence for a linguistic *false-cognate* or *double-entendre* of David Hofstadter fame (Hofstadter

---

[139]   In the *computational* (Gruber 1995) or *Information Science* (Contributors, Ontology (information science) 2021) sense of the term, by means of which the "specification of a conceptualization" is made.

2000) having dual P/meta-P readings. This would not directly contradict the mixed testimony of acquaintances Kreisel, Bernays, Goodstein, and Floyd, immediate and once-removed (Matthíasson 2013), that even early on Wittgenstein had special insight into certain aspects in the actual workings of GI, while coming well wide of the mark on some crucial others. His former student Kreisel even tells of a Wittgenstein who in the 1930s understands Gödel's derivations, though by the 1940s does not.[140]

A reflecting, perhaps brooding Gödel by 1973 was able to speak his mind. He doubted that there had been any confusion whatever on Wittgenstein's part—whether on the reading of g or the meta-linguistic ontology defining "theorem", "proposition", and the like. His suspicion, and perhaps the explanation that best fits the known facts is that Wittgenstein had only pretended, and his "misunderstanding" had been "intentional".[141] Gödel would further openly wonder whether the great philosopher had actually "lost his mind",[142] a suspicion that in later years he would repeatedly express.[143]

As soon after publication of reference (Wittgenstein, *Remarks on the Foundations of Mathematics* 1978) as 1958, Alan Ross Anderson's *Mathematics and the 'Language Game'* (A. Anderson 1958)[144] channels something of this suspicion. Shanker is to be commended for providing fairly comprehensive quotations from a number of cited references authored by mathematicians with views that sharply oppose his own, one here that "reveals Anderson's lack of receptivity":

> Wittgenstein proposes to shed general light on an important locution in metamathematics:
>> However queer it sounds, my task as far as concerns Gödel's proof seems merely to consist in making clear what such a proposition as: "Suppose this could be proved" means in mathematics, (p. 177)

---

[140]   Though just "How plausible that is, is hard to see" (Matthíasson 2013).

[141]   Which, by the way, explains both the extensive testimony to Wittgenstein's understanding of GI and his own words that often suggested misunderstanding.

[142]   Gödel in 1972: "Has Wittgenstein lost his mind? Does he mean it seriously? He intentionally utters trivially nonsensical statements" (Wang 1996, p 179).

[143]   E.g., in letter to Menger "or pretended not to understand it" (Wang, Reflections on Kurt Gödel 1987, p 49). When we consider that in 1953 this same Wittgenstein had written "Philosophy is a battle against the bewitchment of our intelligence by means of language" (Wittgenstein 1953, §109), reluctance to contemplate such a chilling prospect is understandable.

[144]   The title itself a mischievous if not sinister play on the semiotic term made famous by Wittgenstein.

But we are in point of clarity worse off after Wittgenstein's account than before.

There's more. Anderson conducts a close, near line-by-line analysis that finds the "Notorious Paragraph" marred by specious argumentation and its author happy to deploy personal "belittlement", "abusive ad-homonym", and like vain rhetorical devices.[145] And while, "Wittgenstein says [protests that] he feels no antipathy toward these studies themselves, but only toward certain philosophical attitudes associated with them," it is hard to ignore in so much of what Wittgenstein has written what is clear hostility towards the whole project, if not toward the up-and-coming Gödel himself. "Hard words to use of a major philosopher" indeed (Anderson 1958, 451), but one has to wonder (with Gödel) what other plausible explanation might there be?

Anderson, the diplomat, concludes that the "principle trouble" here is that Wittgenstein mistakes Gödel's language-game for another, and that his *Remarks* in particular are "directed to the [language game] battles of twenty-five to fifty years ago" —whose outcomes as happens had largely been determined on the Incompleteness work of Gödel. And while the battles had gone against *those Wittgenstein took to be his opponents* (Frege, Russell, Hilbert and the like) they had not been decided particularly in his favor, that is, for the right reasons (to wit, his own). And so, why did Wittgenstein not see (indeed, "seek out") Gödel as an ally? The avowed reason he makes all too clear. "Philosophical questions" such as those of mathematical Consistency or Completeness, "cannot be answered mathematically" (Shanker 1988).[146]

Anderson sets to clear up this unfortunate "principle trouble" with a patient if stern *reminder* of the unequivocal conceptual context and concomitant language-game in which Gödel's work does appear, the reason for the question it addresses, and GI's answer to it:

---

[145]   "Most people don't understand anything, [and so they don't wonder about anything. I and so they are unable to wonder about anything.] [not about Cantor, Gödel, etc.]" (Wittgenstein 2001, Manuscript 122).

[146]   And must, that is to say, be left to the grown-ups: philosophers. This was to be Wittgenstein's issue alone, his fight, and Gödel's standard interpretation of Incompleteness was pure and simple mathematical meddling into the affairs of philosophy.... In fact, The Logicist and Formalist mathematical programs were at the time coming under increasing scrutiny, though not, as Wittgenstein would have had it, because they admit meaningless sentences or try to solve consistency problems that do not exist. But rather because of the perceived Incompleteness of all axiomatic systems, and the intractability of Hilbert's consistency problem. Neither issue has Gödel simply dismissed or disparaged, but set upon their resolution real, tangible, mathematical limits.

What is to be made of the claim that "True in Russell's system' means: proved[11] in Russell's system"? For one thing, "true", in the context of mathematical logic, is a semantical predicate, and "proved" is syntactical. Given (a) a syntactical formulation of a system, under which certain expressions are provable, and (b) an interpretation of the system, under which certain expressions are true, we may wish to know whether all the truths are provable. If so, the syntactical system is said to be complete relatively to the interpretation; otherwise incomplete. The rough content of Gödel's theorem is that if the truths are those of elementary arithmetic, and if provability is defined for a constructive and consistent formal system, then not all true statements are provable in that system; in particular, PM is incomplete in the sense described. Questions about completeness arise quite naturally in the course of studying formal systems; such questions are in fact very close to the heart of mathematical logic. But Wittgenstein seems to want to legislate this class of questions out of existence, and by peculiarly un-Wittgensteinian methods at that. For surely such use as the phrase "true in Russell's system" has in the literature does not justify the remark under discussion. And it is in any event difficult to reconcile the comment with the claim that he is not saying that the proof "shews something unimportant." If the proof does not show that truth outruns provability in PM (provided the system is consistent), then what of importance does it show?

The matter could not be put more plainly.

Gödel the Mathematician

Vienna to Konigsberg

It is interesting that on a blog-site listing of philosophical questions headed "Why is there something rather than nothing?", the question, "What are numbers?" ranks eighth among those that readers would most like to have answered, two places ahead of "What is the meaning and purpose of life?" (Lipps 2015). The practicing mathematician might be puzzled by such a question, or might flippantly answer "quantity?", but when consulting the professional mathematics journals they find "What are arithmetical truths about?" (Kolman 2006, 45), and in David Potters own *Reason Nearest Kin* "...what then are these [arithmetical] propositions about?" (Potter, Reason's *Nearest Kin: Philosophies of Arithmetic from Kant to Carnap* 2000, 17) they might think again. The question is not about the meaning or use of the noun "number", but is an inquiry into the fundamental reality of the entity it designates.[147] It is a philosophical question: What are numbers, *really*?

---

[147]   When we have realized the obstacles in the way of a straightforward and confident answer, we shall be well launched on the study of philosophy—for philosophy is

Or do they even exist, another good one? For the Greeks they were forms; for Kant, rules for representing quantity; in Frege's world, extensions of concepts, "number" being the number of a class, by a slight extension (Šikić 1996), etc., and every few generations the perennial blogger would plead the question anew, unsatisfied with the received answer. The difficulty in coming to anything reliable here, and the resulting recrudescence, might be indicative of the profound depth of the inquiry. Alternatively, its pervasiveness across a range of questions treated by philosophers might lead one to the more cynical view that not only are they difficult, but that they may very well be intractable, "open", as a matter of practical definition (Floridi 2013, 200). Blogger Buchholtz not unreasonably proposes (Buchholtz 2021).

> "Philosophies" are "thought frameworks". At best, they provide abstractions on "possible answers" or tools for bounding and exploring "possible answers". Their realm is rooted in the "unanswerable". Science is the realm of the answerable.
>
> What science can't answer becomes Philosophy. What Philosophy can't bound becomes Religion.

Thereupon, philosophy may simply dismiss a given question as nonsense or approach it as a miscommunication from either the speaker or listener: either the speaker lacks sufficient linguistic resources to express his genuinely coherent thought, or the listener lacks sufficient understanding of the speaker's language. These are the considerations that in context of the Linguistic Turn in Philosophy around the turn of the century (Glock and Kalhat 2018)[148] led, also in turn, to the Ideal-Language and Ordinary-Language strategies for resolution of the philosophical question.

In the first approach, briefly, the speaker's question is either dismissed outright as nonsense or closed by un-ambiguation of its terms, re-set in a suitable *conceptual framework* (Sellars 1965) and expressed in an Ideal Language, the philosopher playing the part of *conceptual designer* (Floridi 2013, 211). In the second, the listener, having mistaken for non-

---

merely the attempt to answer such ultimate questions, not carelessly and dogmatically, as we do in ordinary life and even in the sciences, but critically, after exploring all that makes such questions puzzling, and after realizing all the vagueness and confusion that underlie our ordinary ideas (B. Russell 1997, 7).

[148]   Whose motto may well have been "The limits of my Language means the limits of my world" (Wittgenstein, Tractatus Logico-Philosophicus 1922, 74).

sense what was not, does a more thorough study of the ordinary language in which the speaker has put the question, a *conceptual analysis* (Thomasson, Ordinary Objects 2007) (Thomasson, Existence Questions 2008), and in the corresponding ordinary language framework, re-discovers the question, closed.

While Wittgenstein's conviction that philosophical problems arise from linguistic confusion remains firm throughout his lifetime, at separate times in its course he advocates both strategies.

In practice, the Ideal Language approach might be augmented by an application rule which sets the degree to which one should take seriously its ontological claims. This might be an observational constraint for ideal-language proper-name referencing—an application condition involving, for example, W. Quine's criterion for the ontological commitment of an Ideal statement that, "A theory is committed to those and only those entities to which the bound variables of the theory must be capable of referring in order that the affirmations made in the theory be true" (Quine 1948, 33).

Or one might instead with R. Carnap forgo such a rule by modeling one's ontological commitments on those made in the practice of science, which avails itself to a second, *instrumental* Ideal Language whose objects are theoretically related to those of the observational Ideal Language, though to which science has no like commitments. Thus, while in the internal conceptual framework and language of some theory, there is a proton now passing through my detector, and the theory tells me that in the everyday external framework in which my detector lives I should hear a beep, which I do, this yet does not commit me to the existence of protons (as some competing theory in which "protons" do not appear may predict the beep just as reliably, or better), which I may take instrumentally (much as infinities are read in Hilbert's finite arithmetic) granting no Ontological Commitment to the terms Ideally referenced (Carnap, Empiricism, semantics, and ontology 1950, 109).

On the question of the existence of abstract objects in particular, Quine (with input from Putnam) offers the criteria, [149] "We ought to have ontological commitment to all and only the entities that are indispensable to our best scientific theories," by which presumed indispensability of

---

[149]   As formulated in reference (Colyvan 2019)

numbers to the conduct of science we may take to condone an observational reading of ideal-language number existence claims—indeed, as an observation *itself* of numbers.

Of course, there could be no question of a like justification in Carnap's scheme, in which any inquiry into some external observational existence of internal instrumental numbers would be devoid of cognitive sense (Carnap, Empiricism, semantics, and ontology 1950, 109). Notice however that there are no substantive disagreements here between the two Ideal Language schemes in respect of numbers (e.g., via their mathematical relations), only different referencing rules.[150]

Nonetheless, the stage had been set, if not for the clarification of abstract object existence claims, then for a Positivist battle for influence over the Ideal Language approach to the open question. In a little-known skirmish in the ensuing Carnap/Quine Controversy, forgettable except for its display of weaponry and tactics, Alonzo Church (citing Quine's criteria) taunts those who like A. J. Ayer would like to assert that "1729 is the sum of two cubes" while denying "1729" ontological status (Church 1958, 1008)

$$F(a) \cap \neg\exists x.F(x)$$

For if as Frege thought, singular linguistic terms *name* real objects (Frege 1884, 73) then surely, with Quine, *to exist* is to be the value of a bound-variable (W. Quine 1948), $F(a) \Rightarrow \exists x.F(x)$, drawing the ideal language contradiction.[151]

Meanwhile the real-world pointlessness of this and like "pseudo-dispute" exercises (Thomasson 2008, 452) did not go unnoticed, and in due course, the rumor would get around the stadium, to borrow again from Thomasson, that the fights aren't real. Amie Thomasson with P. Strawson takes seriously the usefulness and feasibility of the traditional ontological project, for which she thinks Ordinary Languages most suitable for conducting "substantive factual debates about the world"

---

[150]   N. Malcolm notes the same in respect of philosophical disputes as early as 1942 (Malcolm 1942).

[151]   Of course, leaving out of account that nominalist Ayer generally respects Carnap's internal/external statement distinction, for which there is no such contradiction.

(Thomasson 2008, 467), only that its ontological claims must be thoroughly disambiguated by stringent *conceptual analysis* of the basic conditions for use, i.e., for "application", of its nominative terms.[152]

According to Thomasson, in existence claims the need for *semantic descent* among linguistic users enters twice: in agreeing on some correlation between "existence" and "reference" claims. For example, (Thomasson 2008, 67)

E: '2' exist iff *2* refers

(where *2* is the meaning of '2' as e.g. given by some definitional criteria)
and for referencing

R: *2* refers if and only if the application conditions for *2* are fulfilled.

in agreeing on the *conceptually relevant* application conditions: the rules that restrict referencing by "2" except to entities satisfying the criteria. Thus, the criteria itself defines an ontological category to which *2* by definition belongs, and *as such* exists. [153] Church may indeed be in dispute with Ayer over the existence of numbers, though not via Ideal-language Existential Generalization by singular predication (Contributors 2022) in anything that Ayer says, but from conflicting application conditions for "number" objects, such as Quine's criteria, which Church nowhere mentions. In any case, these leaves "the metaphysician work to elucidate the application conditions for our terms," though it is difficult to see how linguists might not have equal right to it. Leaving aside the question of the existence of abstract objects such as numbers, however, Thomasson usually contends at the opposite extreme with Eliminativists who would deny the existence of ordinary tables and chairs.

---

[152] Which become the relevant frame-level application conditions and co-application conditions—"'frame-level' since they involve conditions that are *conceptually* relevant to whether or not reference is established, not all the conditions that may be empirically discovered as relevant" with coapplication conditions as "rules that (supposing the term to have been successfully applied) specify under what conditions the term would be applied again to one and the same entity..." (Thomasson, Ordinary Objects 2007, 39-40).

[153] Though whether Thomasson agrees that Quine's criteria corresponds to "norms of use" for the "conceptual content" of "2" collectively "established by competent speakers of the language" (Thomasson, Ordinary Objects 2007, 29&180) would be another question entirely.

Decades before Thomasson, Peter Strawson argued persuasively for the suitability of conceptual frames corresponding to Ordinary Languages for addressing ontological questions and the like, frames that naturally "lay bare" their own ontological commitments, and also concludes that what is needed is an adequately thoroughgoing conceptual analysis of ordinary use, and not new artificial conceptual schemes (Strawson 1959) (Pelletier 2011, 2). He argues for a metaphysics that only *describes* metaphysical categories in ordinary language as used, and by which such categories obtain their meaning. This would be the subject of Strawson's Descriptive Metaphysics at a time when nominalist thought dominated ontology.

As noted, it is no accident, though remarkable all the same, how closely both conceptions of the philosophical question, from the unanswerably nonsensical to the conceptually under-analyzed and their corresponding strategies, are mirrored in the evolution of Wittgenstein's own philosophy of language—the Ideal of the Tractatus to the Ordinary of Philosophical Investigations, Wittgenstein being both student of Logicist Frege and patron saint of the loosely associated Ordinary-Language school of philosophy. While his late views on Mathematical Platonism and Algorithmic Decidability established by the time of his writing *Remarks on the Foundations of Mathematics* (Wittgenstein 1978) are well known, the fact is almost irrelevant to what would be his actual approach to GI. What Wittgenstein does understand, and knows very well, is the framework contextuality of the language games current at the time, and that in the Formalist framework in which GI appears, his objection to the Gödel sentence g as a "meaningless concatenation of symbols" simply does not hold.[154] Anderson's above exasperation is understandable, for one can hardly think Wittgenstein's linguistic mixing accidental; Gödel certainly didn't (cf. footnote (139)). If there is one complaint against his personal behavior in this regard that may irrevocably stick, that is it.

> We mind about the kind of expressions we use concerning these things [i.e., 'movements of a machine' as deductions in a formal system]; we do not understand them, however, but misinterpret them. When we do philosophy [here, in the traditional sense] we are like savages, primitive people, who hear the expressions of

---

[154]   On the other hand, Pidgen observes in *Coercive Theories of Meaning* that rhetorical nonsense-ascriptions nonetheless function as effective coercive devices, and cites Wittgenstein as one of its worst offenders (Pigden 2010, 165)

civilized men, put a false interpretation on them, and then draw the queerest conclusions from it (Wittgenstein 1953, 194).

Wittgenstein might have done himself a service to have shown some minimum admiration for Gödel's mathematical feat, but even in his Nachlass one finds no trace of it. The mathematician who is not impressed by the Ingenious acrobatics of the derivation cannot be imagined, and the initial reviewers of REM were all confessed mathematicians.

Does Gödel then address a philosophical problem as Wittgenstein insists?

"I am a man," says Huxley (quoting Lawrence) "therefore a novelist," (Huxley 1958, v) to which extent perhaps every man has a right to philosophy. And even as many must have it that, "When posing the old question 'What are arithmetical truths about?'('What is their epistemic status?' or 'How are they possible?') we find ourselves standing in the shadow of Gödel, just as our predecessors stood in the shadow of Kant" (Kolman 2006, 45), Gödel of the Incompleteness years (i.e., of the late 1920s and early 1930s) really shows no interest in such questions. He is content to work out structural details of the prevailing Hilbertian conceptual framework of the day in which his advisor Hahn had placed him, even accounting for his oft-cited attendance at the Vienna Circle meetings (to which advisor Hahn had invited him!) that after all would take place just down the hallway in the very Physics building where he regularly studied.[155] Yet Einstein is noted to have praised Gödel as the foremost philosopher-logician since Aristotle (Budiansky 2021, 178), and others as the greatest of all time (Yourgrau 2006, 1).

Yes, perhaps. We rather picture him grinding away at his usual University of Vienna Strudlhofgasse-4 3E63 study desk, no more than 40 paces around the corner from Boltzmanngasse 5 where the celebrated Circle of philosophers conducted their own quite different business. He is the young mathematician who in the space of 25 condensed pages would produce a proof of the inadequacy of axiomatic number theory so deci-

---

[155]   If leaving aside a 1948 defense of mathematical Platonism (Gödel, What is Cantor's continuum problem? 1947), a forgettable Ontological Argument for the Divine (Gödel, Kurt Gödel: Collected Works: Volume III: Unpublished Essays and Lectures 1986, 429), with a couple like speculative essays written in the 60's, well into his declining years.

sive that for generations of popular literature it would be misappropri-ated to prove also the existence of numbers (cf. "Lisbon & Pisa, 2014"), the transcendence of truth (cf. Ohio, 1998), the non-formalization of hu-man intelligence (cf. "Indirect Self-Reference"), a dozen like speculations and the negation of each. He is the lone gunman who single-handedly changed the course of 20th century critical thinking, the gifted 22-year old mathematics student, Kurt Gödel.

References

Anderson, A, and N Belnap. 1961. "Enthymemes." *Journal of Philosophy* 58 (23): 713-723.

Anderson, A. 1958. "Mathematics and the "Language Game"." *The Review of Metaphysics* 11 (3): 446-458.

Armour-Garb, B. 2004. "Minimalism, the Generalization Problem and the Liar." *Synthese* 139: 491-512.

Askanas, M. 2006. *Godel's Incompleteness Theorems—A Brief Introduction.* Accessed March 9, 2021. http://math.mind-crafts.com/godels_incompleteness_theorems.php.

Azzouni, J. 1999. "Comments on Shapiro." *The Journal of Philosophy* 96 (10): 541-544.

—. 2006. *Tracking Reason: Proof, Consequences, and Truth.* Oxford University Press.

Bar-Hillel, Y. 1954. "Logical Syntax and Semantics." *Language* 30 (2): 230-237.

Beklemishev, L., and S. Artemov. 2004. "Provability Logic." *Handbook of Philosophical Logic, Second Edition* 13: 229-403.

Belnap, N. 2006. "Prosentence, Revision, Truth, and Paradox." *Philosophy and Phenomenological Research* LXXIII (3): 705-712.

Benacerraf, P. 1981. "Frege: The Last Logicist." *Midwest Studies in Philosophy* 6 (1): 17-36.

Benveniste, E. 1971. *Problems in General Linguistics, Issue 8 of Miami linguistics series.* University of Miami Press.

Berto, F. 2009. "The Gödel Paradox and Wittgenstein's Reasons." *Philosophis Mathematica* 17 (2): 208-219.

Beziau, J.Y. 2007. *Logica Universalis: Towards a General Theory of Logic.* Springer.

Blanchette, P. 1996. "Frege and Hilbert on Consistency." *The Journal of Philosophy* 93 (7): 317-336.

Blanchette, P. 2017. "Models in geometry and logic: 1870-1920." *n S. S. Niniil-Logic, methodology and philosophy of science: Proceedings of the 15th international congress* 41-61.

Blanchette, P. 2018. "The Frege-Hilbert Controversy." In *The Stanford Encyclopedia of Philosophy*, edited by Edward N. Zalta. Metaphysics Research Lab, Stanford University. Accessed March 8, 2021. https://plato.stanford.edu/archives/fall2018/entries/frege-hilbert.

Blanck, R. 2017. *Contributions to the Metamathematics of Arithmetic. Fixed points, Independence, and Flexibility.* PhD thesis: University of Gothenburg.

Blasjo, V. 2008. *Gödel's Theorem in Focus (Philosophers in Focus)*. February 8. https://www.amazon.com/Gödel's-Theorem-Focus-Philosophers/dp/041 5045754.

Boolos, G. 1994. "Gödel's Second Incompleteness Theorem Explained in Words of One Syllable." *Mind* 103 (409): 1-3.

Boolos, G, J. Burgess, and R Jeffrey. 2002. *Computability and Logic*. Cambridge University Press.

Bridges, D., and E. Palmgren. 2018. "Constructive Mathematics." In *The Stanford Encyclopedia of Philosophy*, edited by Edward N. Zalta. Metaphysics Research Lab, Stanford University. https://plato.stanford.edu/archives/sum 2018/entries/mathematics-constructive.

Buchholtz, R. 2021. *Quora: To what questions do philosophers not yet have an answer*. June 26. https://www.quora.com/To-what-questions-do-philosophers-not-yet-have-an-answer.

Budiansky, S. 2021. *Journey to the Edge of Reason: The Life of Kurt Gödel*. Oxford University press.

Buldt, B. 2014. "The Scope of Godel's First Incompleteness." *Logica Universalis* 8 (3-4): 499-552.

Bundy, A., M. Atiyah, A. Macintyre, and D. MacKenzie. 2005. "The nature of mathematical proof." *Philosophical Transactions of the Royal Society A* 363 (1835): 2329-2461.

Butrick, R. 1965. "The Godel Formula: Some Reservations." *Mind* 74 (295): 411-414.

Cantini, A. 2009. "Paradoxes, self-reference and truth in the 20th century." In *Handbook of the history of logic (vol. 5)*, 875-1013. Amsterdam: Elsevier/North-Holland.

Carnap, R. 1950. "Empiricism, semantics, and ontology." *Revue Internationale de Philosophie* 4 (11): 20-40.

—. 1934. *The Antinomies and the Incompleteness of Mathematics*. Monthly for Mathematics and Physics.

—. 1934. *The Logical Syntax of Language*. K. Paul, Trench, Trubner & Company, Limited.

Cellucci, C. 2009. "The universal generalization problem." *Logique & Analyse* 52: 3-20.

Chalmers, D. 1995. "FACING UP TO THE PROBLEM OF CONSCIOUSNESS." *Journal of Consciousness Studies* 2 (3): 200-19.

Church, A. 1958. "Symposium: Ontological Commitment." *Journal of Philosophy* 1008-1014.

Cieśliński, C. 2010. "Truth, Conservativeness, and Provability." *Mind* 119 (474): 409-422.

Colyvan, M. 2019. "Indispensability Arguments in the Philosophy of Mathematics." In *The Stanford Encyclopedia of Philosophy*. Metaphysics Research Lab, Stanford University. https://plato.stanford.edu/archives/spr2019/entries/mathphil-indis/.

Contributors, Wikipedia. 2022. *Existential Generalization*. Wikipedia, the free encyclopedia. Accessed December 23, 2022. https://en.wikipedia.org/wiki/Existential_generalization.

—. 2021. *False Equivalence*. Wikipedia, The Free Encyclopedia. Accessed November 11, 2021. https://en.wikipedia.org/w/index.php?title=False_equivalence&oldid=1051100868.

—. 2021. *False friend*. Wikipedia, The Free Encyclopedia. Accessed March 9, 2021. https://en.wikipedia.org/w/index.php?title=False_friend&oldid=1010294429.

—. 2021. *Fundamental Theorem of Arithmetic*. Wikipedia, The Free Encyclopedia. Accessed March 9, 2021. https://en.wikipedia.org/w/index.php?title=Fundamental_theorem_of_arithmetic&oldid=1010813708.

—. 2021. *Gödel's incompleteness theorems*. Wikipedia, The Free Encyclopedia. Accessed March 9, 2021. https://en.wikipedia.org/w/index.php?title=G%C3%B6del%27s_incompleteness_theorems&oldid=1011025054.

—. 2021. *Ignoramus et ignorabims*. Wikipedia, The Free Encyclopedia. Accessed March 9, 2021. https://en.wikipedia.org/w/index.php?title=Ignoramus_et_ignorabimus&oldid=1003182304.

—. 2021. *Metoymy*. Wikipedia, The Free Encyclopedia. Accessed November 11, 2021. https://en.wikipedia.org/w/index.php?title=Metonymy&oldid=1054264519.

—. 2021. *Ontology (information science)*. Wikipedia, The Free Encyclopedia. Accessed November 11, 2021. https://en.wikipedia.org/w/index.php?title=Ontology_(information_science)&oldid=1054529999.

—. 2021. *Pyrrhon of Elis*. Wikipedia, The Free Encyclopedia. Accessed November 11, 2021. https://en.wikipedia.org/w/index.php?title=Pyrrhon_of_Elis&oldid=17622636.

Correia, F. 2004. "Semantics for Analytic Containment." *Studia Logica* 77 (1): 87-104.

Curry, H. 1968. "THE PURPOSES OF LOGICAL FORMALIZATION." *Logique et Analyse, NOUVELLE SÉRIE* 11 (43): 357-366.

David, M. 2002. "Minimalism and the Facts about Truth." In *What is Truth?*, 161-175. Berlin: Walter de Gruyter.

Davidson, D. 1967. "Truth and Meaning." *Synthese* 17 (3): 304-323.

Dawson, J. 1985. "Completing the Godel-Zermelo Correspondence." *Historia Mathematica* 12: 66-70.

Dawson, J. 1984. "Discussion on the foundation of mathematics." *History and Philosophy of Logic* 5 (1): 111-129.

—. 2005. *Logical Dilemmas.* Taylor & Francis.

Dawson, J. 1988. "The Reception of Gödel's Incompleteness Theorems." In *Gödel's Theorem in Focus*, by S. Shanker, 74-95. New York: Croom Helm.

Dawson, J. 1997. "The Reception of Godel's Incompleteness Theorems." In *Godel's Theorem in Focus*, 74-96. New York: Routledge.

Detlefsen, M. 1979. "On interpreting Gödel's Second Theorem." *Journal of Philosophical Logic* 8: 297-313.

Dummett, M. 1963. "The Philosophical Significance of Gödei's Theorem." *Ratio* 5: 140-155.

Einstein, A, B Podolsky, and N Rosen. 1935. "Can Quantum-Mechanical Description of Physical Reality be Considered Complete?" *Physical Review* 47 (10): 777-780.

Feferman, S. 1998. "Deciding the Undecidable." In *In Light of Logic*, 3-28. New York: Oxford University Press.

Feferman, S. 1986. "Introductory note to 1931c." In *Kurt Godel: Collected works, vol I*, 208-213. London: Oxford University Press.

Feferman, S. 1997. "Kurt Godel: Conviction and Caution." In *Godel's Theorem in Focus*, 96-115. New York: Routledge.

Feferman, S. 2006. "The impact of the incompleteness theorems on mathematics." *Notices Amer. Math. Soc.* 53: 434-439.

Feferman, S. 1962. "Transfinite recursive progressions of axiomatic theories." *The Journal of Symbolic Logic* 27: 259-316.

Ferreira, J. 2008. *On the Goedel's formula.* arxiv.org. Accessed March 9, 2021. https://arxiv.org/pdf/math/0104025.pdf.

Field, H. 1999. "Deflating the Conservativeness Argument." *The Journal of Philosophy,* 96 (10): 533-540.

Field, H. 1994. "Disquotational Truth and Factually Defective Discourse." *The Philosophical Review* 103 (3): 405-452.

Field, H. 1972. "Tarski's Theory of Truth." *The Journal of Philosophy* 69 (13): 347-375.

Fitting, M. 2020. "Intensional Logic." In *The Stanford Encyclopedia of Philosophy*, edited by Edward N. Zalta. Metaphysics Research Lab, Stanford University. Accessed March 9, 2021. https://plato.stanford.edu/archives/spr2020/entries/logic-intensional.

FitzPatrick, F.J. 1966. "To Godel via Babel." *Mind* 75 (299): 332-350.

Floridi, L. 2013. "What is a philosophical question?" *Metaphilosophy* 44 (3): 195-221.

Floyd, J,, and H. Putnam. 2000. "A Note on Wittgenstein's "Notorious Paragraph" about the Gödel Theorem." *The Journal of Philosophy* 97 (11): 624-632.

Floyd, J. 2001. "Prose versus Proof: Wittgenstein on Godel, Tarski and Truth." *Philosophica Mathematica* 3 (9): 280-307.

Frege, G. 1967. *Basic Laws of Arithmetic: Exposition of the System.* Edited by Montgomery Furth. University of California Press.

—. 1879. *Begriffsschrift, a formula language modelled upon arithmetic.* Halle

Frege, G. 1948. "Sense and Reference." *The Philosophical Review* 57 (3): 209-230.

—. 1884. *The Foundations of Arithmetic.*

—. 1918. *Thoughts.* Oxford: Blackwell.

Gödel, K. 1986. *Kurt Gödel: Collected Works: Volume III: Unpublished Essays and Lectures.* Oxford University Press.

—. 1962. *On Formally Undecidable Propositions of Principia Mathematica and Related Systems.* Translated by B. Meltzer. New York: Dover Publications, Inc. http://www.geier.hu/GOEDEL/Godel_orig/godel2.htm.

—. 2000. *On Formally Undecidable Propositions of Principia Mathematica and Related Systems.* Translated by Martin Hirzel. Dover Publishing, Inc. https://drive.google.com/file/d/1q0mT_SSoZssiFqYJInE38ajTFQL1xl2e/view?usp=sharing.

Gödel, K. 1947. "What is Cantor's continuum problem?" *The American Mathematical Monthly* 54 (9): 515-525.

Gödel, K., and S. Feferman. 1986. *Kurt Gödel: Collected Works: Volume III: Unpublished Essays and Lectures.* Kiribati: OUP USA.

Gauker, C. 2001. "T-Schema Deflationism versus Gödel's First Incompleteness Theorem." *Analysis* 61 (2): 129-136.

Girard, J.Y. 2011. *The Blind Spot: Lectures on Logic.* Zurich: European Mathematical Society.

Glanzberg, M. 2018. "Truth/1.1.1 The origins of the correspondence theory." In *The Stanford Encyclopedia of Philosophy*, edited by Edward N. Zalta. Metaphysics Research Lab, Stanford University. https://plato.stanford.edu/archives/fall2018/entries/truth/#OriCorThe.

Glanzberg, M. 2018. "Truth/2.4 Reference and Satisfaction." In *The Stanford Encyclopedia of Philosophy*, edited by Edward N. Zalta. Metaphysics Research Lab, Stanford University. https://plato.stanford.edu/entries/truth/#RefSat.

Glanzberg, M. 2018. "Truth/5.1 The Redundancy Theory." In *The Stanford Encyclopedia of Philosophy*, edited by Edward N. Zalta. https://plato.stanford.edu/entries/truth/#RedThe.

Glanzberg, M. 2018. "Truth/5.2 Minimalist Theories." In *The Stanford Encyclopedia of Philosophy*, edited by Edward N. Zalta. Metaphysics Research Lab, Stanford University. https://plato.stanford.edu/archives/fall2018/entries/truth/#MinThe.

Glock, H., and J. Kalhat. 2018. "Linguistic turn." In *The Routledge Encyclopedia of Philosophy*. Taylor and Francis. https://www.rep.routledge.com/articles/thematic/linguistic-turn/v-1.

Grover, D, J Camp, and N Belnap. 1975. "A Prosentential Theory of Truth." *Philosophical Studies* 27 (2): 73-125.

Gruber, T. 1995. "Toward Principles for the Design of Ontologies Used for Knowledge Sharing." *International Journal of Human-Computer Studies* 43 (5-6): 907-928.

Gómez-Torrente, M. 2019. "Alfred Tarski." In *The Stanford Encyclopedia of Philosophy*, edited by Edward N. Zalta. Metaphysics Research Lab, Stanford University. https://plato.stanford.edu/archives/spr2019/entries/tarski.

Gupta, A. 1993. "A Critique of Deflationism." *Philosophical Topics* 21 (2): 57-81.

Haase, C. 2018. "A survival guide to Preburger Arithmetic." *ACM SIGLOG News* 5 (3): 67-82.

Halbach, V. 2001. "Disquotational Truth and Analyticity." *Journal of Symbolic Logic* 66: 1959-1973.

Halbach, V. 1999. "Disquotationalism and Infinite Conjunctions." *Mind* 108 (429): 1-22.

Halbach, V. 2001. "How Innocent Is Deflationism?" *Synthese* 126 (1/2): 167-194.

Haskel, M. 2015. *A Brief Taste of Quantifiers.* University of Notre Dame. August 22. https://www3.nd.edu/~mhaskel/talks/quantifiers-notes.pdf.

Helmer, O. 1937. "Perelman versus Godel." *Mind* 46 (181): 58-60.

Hendricks, V., K. Jorgensen, and S. Pedersen. 2013. *Knowledge Contributors, Volume 322 of Synthese Library.* Springer Science & Business Media.

Herbrand, J. 1932. "Sur la non-contradiction de l'Arithmetique." *Journal fur die reine und angewandte Mathematic* 1932 (166): 1-8.

Hodges, W. 2018. "Tarski's Truth Definitions." In *The Stanford Encyclopedia of Philosophy*, edited by Edward N. Zalta. Metaphysics Research Lab, Stanford University. Accessed March 9, 2021. https://plato.stanford.edu/archives/fall2018/entries/tarski-truth/#ObjLanMet.

Hofstadter, D. 2000. *Gödel, Escher, Bach: an Eternal Golden Braid*. Penguin Books.

Hofstadter, D. 1999. "Mathematician Kurt Godel." *TIME* 153 (12).

Hoge, S. 2010. *New essay on Goedel's Incompleteness Theorem.* Accessed March 9, 2021. http://sci.logic.narkive.com/ruLx38b7/new-essay-on-goedel-s-incompleteness-theorem.

—. 2010. *Supernatural Numbers And Undecidable Arithmetical Statements.* https://web.archive.org/web/20100108063237/http://www.hoge-essays.com/incompleteness.html.

Horrigan, P. 2014. *Transcendental Truth.* https://www.academia.edu/20120297/Transcendental_Truth: Academia.edu.

Horsten, L. 2011. *The Tarskian Turn: Deflationism and Axiomatic Truth.* The MIT Press.

Huxley, A. 1958. *Collected Essays of Aldous Huxley.* Harper & Brothers.

Iemhoff, R. 2020. "Intuitionism in the Philosophy of Mathematics." In *The Stanford Encyclopedia of Philosophy*, edited by Edward N. Zalta. Accessed March 9, 2021. https://plato.stanford.edu/archives/fall2020/entries/intuitionism.

Ignjatović, A. 1994. "Hilbert's Program and the Omega-Rule." *The Journal of Symbolic Logic* 59 (1): 322-343.

Jamnik, M. 2001. *Mathematical reasoning with diagrams: from intuition to automation.* Stanford, CA.: CSLI Press.

Jones, J. 2007. *My investigations into Godel's Incompleteness Theorem.* Accessed March 9, 2021. http://sci.logic.narkive.com/d7q8K6qv/my-investigations-into-godels-incompleteness-theorem.2.

—. 2008. *My talk about Godel to post-grads.* Vers. 12:02:58 PM. Google Groups. June 22. https://groups.google.com/g/sci.logic/c/y0QcsK10s58?pli=1.

Jones, J., and W. Wilson. 2009. *An Incomplete Education.* Random House.

Ketland, J. 2000. "Conservativeness and translation-dependent T-schemes." *Analysis* 60 (4): 319-328.

Ketland, J. 1999. "Deflationism and Tarski's Paradise." *Mind* 108 (429): 69-94.

Ketland, J. 2005. "Deflationism and the Gödel Phenomena: Reply to Tennant." *Mind* 114 (453): 75-88.

Ketland, J. 2010. "Truth, Conservativeness, and Provability: Reply to Cieśliński." *Mind* 119 (474): 423-436.

Kleene, S.C. 1971. *Introduction to Metamathematics.* Groningen: Wolters Noordhoff.

Kohler, E, and J Wolenski. 2013. *Alfred Tarski and the Vienna Circle: Austro-Polish Connections in Logical Empiricism.* Springer Science & Business Media.

Kolman, V. 2006. "Gödel's Theorems and the Synthetic-Analytic Distinction." *Miscellana Logica 6* 45-61.

Krajewski, S. 2004. "Godel on Tarski." *Annals of Pure and Applied Logic* 127 (1-3): 303-323.

Kreisel, G. 1958. "Wittgenstein's Remarks on the Foundations of Mathematics." *The British Journal for the Philosophy of Science* 9 (34): 135-158.

Kreisel, G., and A. Levy. 1968. "Reflection Principles and their Use for Establishing the Complexity of Axiomatic Systems." *Mathematical Logic Quarterly* 14 (7-12): 97-142.

Lacey, H, and G Joseph. 1968. "What the Godel Formula Says." *Mind* 77 (305): 77-83.

Lajevardi, S., and S. Salehi. 2021. "There May Be Many Arithmetical Gödel Sentences." *Philosophia Mathematica* nkaa041. https://doi.org/10.1093/phil mat/nkaa041.

Lampert, T. 2006. "Wittgenstein's 'notorious paragraph' about the Godel Theorem." *Contributions of the Austrian Wittgenstein Society 2006.* Kirchberg am Wechsel.

Leigh, G., and V. Halbach. 2020. "Axiomatic Theories of Truth." In *The Stanford Encyclopedia of Philosophy*, edited by Edward N. Zalta. Accessed March 9, 2021. https://plato.stanford.edu/archives/spr2020/entries/truth-axiomat ic/#CompTrut.

Leitgeb, H. 2001. "Truth as Translation—Part A." *Journal of Philosophical Logic* 30: 281-307.

Lipps, K. 2015. *Quora: What are the top 10 big philosophical questions most people wonder about?* May 12. { https://www.quora.com/What-are-the-top-10-big-philosophical-questions-most-people-wonder-about.

Lipscomb, T. 2010. "Gödel's Incompleteness Theorems: A Revolutionary View of the Nature of Mathematical Pursuits." *Rose-Hulman Undergraduate mathematics Journal* 11 (1): Article 8.

Locke, J. 1975. *An Essay Concerning Human Understanding.* Oxford: Oxford University Press.

Longo, G. 2011. "Reflections on Concrete Incompleteness." *Philosophia Mathematica* 19 (3): 255-280.

Lucas, J.R. 2002. *Conceptual Roots of Mathematics* (Chapter 8). Routledge.

Lucas, J.R. 1961. "Minds, Machines and Godel." *Philosophy* 36 (137): 112-127.

Lucy, J. 1993. "Reflexive language and the human disciplines." In *Reflexive Language: Reported Speech and Metapragmatics*, 9-32. Cambridge: Cambridge University Press.

Lyons, J. 1977. *Semantics: Volume I.* New York: Cambridge University Press.

Makey, J. 1995. *Godel's Incompleteness Theorem is not an Obstacle to Artificial Intelligence.* Rose-Hulman Institute of Technology. Accessed March 9, 2021. https://www.sdsc.edu/~jeff/Godel_vs_AI.html.

Malcolm, N. 1942. "Moore and Ordinary Language." In *Ordinary Language*, by V. Chappell, 5-23. Prentice Hall. https://iep.utm.edu/ord-lang/#SH7e.

Mancosu, P. 1999. "Between Vienna and Berlin: The immediate reception of Godel's incomplenteness theorems." *History and philosophy of logic* 20: 33-45.

Mangin, S. 2002. "Gödel's Incompleteness Theorems." Thesis (Bachelor of Arts), Philosophy, University of Melbourne. https://web.archive.org/web/2014 0410163404/http://sephorahmangin.info/selected_essays/Godel_Incomp leteness_Theorems.pdf.

Marker, D. 2006. "Gödel's Incompleteness Theorem I: Represening Primitive Recursive Functions." *Metamathematics I.* Accessed March 9, 2021. http://homepages.math.uic.edu/~marker/math502-03/meta9.pdf.

Mattessich, R. 2012. *Instrumental Reasoning and Systems Methodology.* Springer Netherlands.

Matthíasson, Á. 2013. "A Chalet on Mount Everest: Interpretations of Wittgenstein's Remarks on Gödel." Thesis (MSc.), Philosophy, Universiteit van Amsterdam. https://eprints.illc.uva.nl/id/eprint/915/1/MoL-2013-26.text.pdf .

McCarthy, T. 2016. "Gödel's Third Incompleteness Theorem." *dialectica* 70 (1): 87-112.

McCarthy, T., and K. Fine. 1984. "Truth without Satisfaction." *Journal of Philosophical Logic* 13 (4): 397- 421.

Moschovakis, Y. 2014. "Lecture Notes in Logic." *Yiannis N. Moschovakis. Books and lectures notes on line.* UCLA Mathematics. March 29. https://www. math.ucla.edu/~ynm/lectures/lnl.pdf.

Mostowski, A. 1952a. "On models of axiomatic systems." *Fundamenta Mathematicae* 39: 133-158.

Murawski, R. 1998. "Undefinability of Truth. The Problem of Priority: Tarski vs Godel." *HISTORY AND PHILOSOPHY OF LOGIC* 19: 153-160.

OED Online. 2022. *Semantics.* Oxford University Press. https://www.oed.com/ view/Entry/345083?redirectedFrom=semantics.

Pelletier, F. 2011. "Descriptive metaphysics, natural language metaphysics, Sapir–Whorf, and all that stuff." *The Baltic international yearbook of cognition, logic and communication 6* 1-46.

Piazza, M., and G. Pulcini. 2015. "A Deflationary Account of the Truth of the Godel Sentence G." In *From Logic to Practice*, 71-90. Springer.

Pigden, C. 2010. "Coercive Theories of Meaning or Why Language Shouldn't Matter (So Much) to Philosophy." *Logique et Analyse* 151-184.

Potter, M. 2000. *Reason's Nearest Kin: Philosophies of Arithmetic from Kant to Carnap.* Oxford University Press.

Potter, M. 2011. "Wittgenstein on Mathematics." In *The Oxford Handbook on Wittgenstein*, by M. McGinn and O. Kuusela, 122-137. Oxford: Oxford University Press.

Presburger, M. 1929. "Über die Vollständigkeit eines gewissen Systems der Arithmetik ganzer Zahlen." *Comptes Rendus du I congres de Mathematiciens des Pays Slaves* 92-101.

Priest, G. 2004. "Wittgenstein's Remarks on Gödel's Theorem." In *Wittgenstein's Lasting Significance*, by M. Kölbel and B. Weiss, 206-225. London: Routledge.

Prochazka, K. 2010. *Truth between Syntax and Semantics*. dissertation thesis, Institute of Philosophy and Religious Studies, Charles University, Univerzita Karlova, Filozofická fakulta. https://dodo.is.cuni.cz/bitstream/handle/20. 500.11956/31571/140013917.pdf?sequence=1&isAllowed=y.

Putnam, H. 1979. *Philosophical Papers: Volume 2, Mind, Language and Reality*. Cambridge University Press.

Quine. 1948. "On What There Is." *The Review of Metaphysics* 21-38.

Quine, W. 1948. "On What There Is." *The Review of Metaphysics* 2 (5): 21-38.

Quine, W.V.O. 1970. *Philosophy of Logic.* Havard University Press.

Raatikainen, P. 2008. "Truth, meaning, and translation." In *New Essays on Tarski and Philosophy*, 247-262. New York: Oxford University Press.

Rathjen, M. 2009. "The Constructive Hilbert Program and the Limits of Martin-Lof Type Theory." In *LOGICISM, INTUITIONISM, AND FORMALISM: WHAT HAS BECOME OF THEM?*, 397-433. Synthese Library.

Rautenburg., W. 2006. *A Concise Introduction to Mathematical Logic*. Springer-Verlag.

Reynolds, P.D. 1971. *A Primer in Theory Construction.* Bobbs-merrill Co.

Robinson, J. 1949. "Definability and Decision Problems in Arithmetic." *Journal of Symbolic Logic* 32 (3): 98-114.

Rodych, V. 1999. "Wittgenstein's Inversion of Gödel's Theorem." *Erkenntnis (1975-)* 51 (2/3): 173-206.

Rodych, V. 2018. "Wittgenstein's Philosophy of Mathematics." In *The Stanford Encyclopedia of Philosophy*, edited by Edward N. Zalta. Metaphysics Research Lab, Stanford University. https://plato.stanford.edu/archives/spr2018/entries/wittgenstein-mathematics.

Rosser, B. 1937. "Gödel Theorems for Non-Constructive Logics." *The Journal of Symbolic Logic* 2 (3): 129-137.

Rota, G. 1991. "The Concept of Mathematical Truth." *The Review of Metaphysics* 44 (3): 483-494.

Rucker, R. 2008. *Infinity and the Mind.* New Age International.

Russell, B. 1997. *The Problems of Philosophy.* Oxford University Press.

Russell, B. 1988. "Mathematical Logic as Based on The Theory of Types." Chap. II in *Logic and Knowledge: Essays 1901-1950*, 57-103. Psychology Press.

Sandu, G., and T. Hyttinen. 2004. "Deflationism and Arithmetical Truth." *dialectica* 58 (3): 413-426.

Schönbaumsfeld, G. 2016. "'Hinge Propositions' and the 'Logical' Exclusion of Doubt." *Schönbaumsfeld, G. (2016). 'Hinge PropoInternational Journal for the Study of Skepticism* 6 (2-3): 165-181.

Schantz, R. 2002. *What is Truth?* Berlin: Walter de Gruyter.

Schindler, T., and L. Picollo. 2019. "Deflationism and the Function of Truth." *Philosophical Perspectives* 32: 326-351.

Schmerl, U. 1982. "Iterated Reflection Principles and the ω-Rule." *The Journal of Symbolic Logic* 47 (4): 721-733.

Sellars, W. 1965. *Scientific realism or Irenic Instrumentalism.* Vol. II, in *Boston Studies in the Philosophy of Science*, by R Cohen and M. Wartofsky, 171-205. Humanities Press Inc.

Shanker, S. 1987. *Wittgenstein and the Turning Point in the Philosophy of Mathematics.* State University of New York Press.

Shanker, S. 1988. "Wittgenstein's Remarks on the Significance of Gödel's Theorems." In *Gödel's Theorem in Focus*, by S. Shanker, 155-256. New York: Croom Helm.

Shapiro, S. 1991. *Foundations without Foundationalism.* Oxford University Press.

Shapiro, S. 1998. "Proof and truth: Through thick and thin." *The Journal of Philosophy* 95 (10): 493-521.

Shavrukov, V. 1997. "Interpreting Reflexive Theories in Finitely Many Axioms." *Fundamenta Mathematicae* 152: 99-116.

Shin, S.J. 2015. "Quantifiers Are Logical Constants, but Only Ambiguously." In *Quantifiers, Quantifiers,Quantifiers: Themes in Logic, Metaphysics, and Language*, 51-73. Springer.

Šikić, Z. 1996. "What are Numbers?" *International Studies in the Philosophy of Science* 159-171.Simmons, K. 1993. *Universality and the Liar.* Cambridge University Press .

Smith, P. 2013. *An Introduction to Gödel's Theorems, second edition.* Cambridge University Press.

Smorynski, C. 2009. "Peter Smith. An Introduction to Godel's Theorems." *Philosophica Mathematica* 18 (1): 122-127. https://academic.oup.com/philm at/article/18/1/122/1482306.

Soames, S. 1999. *Understanding Truth.* Oxford University Press.

Soare, R. 1996. "Computability and Recursion." *Bulletin of Symbolic Logic* 2 (3): 284-321.

—. 2016. *Turing Computability.* Chicago: Springer.

Sokal, A., and J. Bricmont. 1998. *Fashionable Nonesense.* Picador.

Steiner, M. 2001. "Wittgenstein as His Own Worst Enemy: The Case of Gödel's Theorem." *Philosophia Mathematica* 9 (3): 257-279.

Stoljar, D., and N. Damnjanovic. 2014. "The Deflationary Theory of Truth." In *The Stanford Encyclopedia of Philosophy*, edited by Edward N. Zalta. Metaphysics Research Lab, Stanford University. Accessed March 10, 2021. https://plato.stanford.edu/archives/fall2014/entries/truth-deflationary/#UtiDef Tru.

Strawson, P. 1959. *Individuals: An Essay in Descriptive Metaphysics.* Methuen.

Suber, P. 2002. *Godel's Proof.* Earlham College. Accessed March 9, 2021. http://legacy.earlham.edu/~peters/courses/logsys/g-proof.htm.

—. 2002. *Sample Formal System S.* Earlham College. Accessed March 10, 2021. http://legacy.earlham.edu/~peters/courses/logsys/sys-xmpl.htm.

Sutner, K. 2017. *Arithmetical Hierarchy.* Pittsburgh: Carnegie Mellon University.

Tait, W. 1981. "Finitism." *The Journal of Philosophy* 78 (9): 524-546.

Tarski, A. 1933. *The Concept of Truth in Formalized Language.* Nakładem Towarzystwa Naukowego Warszawskiego.

Tarski, A. 1944. "The Semantic Conception of Truth: and the Foundations of Semantics." *Philosophy and Phenomenological Research* 4 (3): 341-376.

Tennant, N. 2010. "Deflationism and the Gödel Phenomena: Reply to Cieśliński." *Mind* 119 (474): 437-450.

Tennant, N. 2005. "Deflationism and the Gödel Phenomena: Reply to Ketland." *Mind* 114 (453): 89-96.

Tennant, N. 2002. "Deflationism and the Godel Phenomena." *Mind* 111: 551-582.

Thomas, DW. 1995. "Godel's Theorem and Postmodern Theory." *PMLA* 110 (2): 248-261.

Thomasson, A. 2008. "Answerable and Unanswerable Questions." In *Metametaphysics*, by D Chalmers, D Manley and R Wasserman, 444-471. Oxford University Press.

Thomasson, A. 2008. "Existence Questions." *Philosophical Studies* 141 (1): 63-78.

—. 2007. *Ordinary Objects.* Oxford University Press.

Turing, A. 1939. "Systems of logic based on ordinals." *Proceedings of the London Mathematical Society* 45 (2): 161-228.

Turing, A.M. 1950. "Computing Machinery and Intelligence." *Mind* 59 (236): 433-460.

Uspenski, V.A. 1989. *Godel's Incompleteness Theorem.* Victor Kamkin.

Vidal-Rosset, J. 2006. "Does Godel's Incompleteness Theorem prove that Truth Transcends Proof?" In *The Age of Alternative Logics: Assessing Philosophy of Logic and Mathematics Today*, 51-73. Springer.

Waismann, F. 2003. *Introduction to Mathematical Thinking.* Courier Dover.

—. 1979. *Wittgenstein and the Vienna Circle.* Oxford: Basil Blackwell.

Walsh, Jill Patson. 2011. *Knowledge of Angels.* Transworld.

Wang, H. 1996. *A Logical Journey: From Gödel to Philosophy.* Cambridge, Mass: MIT Press.

—. 1987. *Reflections on Kurt Gödel.* Cambridge, Mass.: MIT Press.

Whitehead, A., and B. Russell. 1910. *Principia Mathematica.* Cambridge University Press.

Williams, M. 2002. "On Some Critics of Deflationism." In *What is Truth?*, 146-160. Berlin: Walter de Gruyter.

Wittgenstein, L. 1974. *Philosophical Grammar.* Oxford: Basil Blackwell.

—. 1953. *Philosophical Investigations.* Oxford: Blackwell.

—. 1964. *Philosophical Remarks.* Oxford: Blackwell.

—. 1978. *Remarks on the Foundations of Mathematics.* Oxford: Blackwell.

—. 1922. *Tractatus Logico-Philosophicus.* London: Routledge & Kegan Paul.

—. 2001. *Wittgenstein's Nachlass, The Bergen Electronic Edition CD-Rom for Windows.* Oxford: Oxford University Press.

Yaqub, A.M. 2008. "Two Types of Deflationism." *Synthese* 165: 77-106.

Yourgrau, P. 2006. *A World Without Time: The Forgotten Legacy of Gödel and Einstein.* Basic Books.

Zach, R. 2019. "Hilbert's Program/1.1 Early work on foundations." In *The Stanford Encyclopedia of Philosophy*, edited by Edward N. Zalta. Metaphysics Research Lab, Stanford University. Accessed March 10, 2021. https://plato.stanford.edu/archives/fall2019/entries/hilbert-program/#1.1.

Zach, R. 2019. "Hilbert's Program/1.2 The influence of Principia Mathematica." In *The Stanford Encyclopedia of Philosophy*, edited by Edward N. Zalta. Metaphysics Research Lab, Stanford University. Accessed March 10, 2021. https://plato.stanford.edu/archives/fall2019/entries/hilbert-program/#1.2.

Zardini, E. 2015. "∀ and ω." In *Quantifiers, Quantifiers, and Quantifiers: Themes in Logic Metaphysics, and Language*, 489-526. Springer.

Zermelo, E. 1932. "Über Stufen der Quantifikation und die Logik." *Jahresbericht der deutschen Mathematiker-Vereinigung* 41 (part 2): 85-88.

Zygmunt, J. 1991. "Mojżesz presburger: life and work." *History and Philosophy of Logic* 12 (2): 211-223.